UNDERSTANDING INSTITUTIONS

UNDERSTANDING INSTITUTIONS

The Science and Philosophy of Living Together

Francesco Guala

PRINCETON UNIVERSITY PRESS
PRINCETON AND OXFORD

Published by Princeton University Press, 41 William Street,
Princeton, New Jersey 08540
In the United Kingdom: Princeton University Press,
99 Banbury Road, Oxford OX2 6JX

press.princeton.edu

First paperback printing, 2023
Paperback ISBN 9780691242354

The Library of Congress has cataloged the cloth edition as follows:

Names: Guala, Francesco, 1970– author.
Title: Understanding institutions : the science and philosophy of living together / Francesco Guala.
Description: Princeton, NJ : Princeton University Press, 2016. | Includes bibliographical references and index.
Identifiers: LCCN 2016004753 | ISBN 9780691171784 (hardcover)
Subjects: LCSH: Social institutions. | Social interaction.
Classification: LCC HM826 .G815 2016 | DDC 306—dc23 LC record available at http://lccn.loc.gov/2016004753

British Library Cataloging-in-Publication Data is available

This book has been composed in Linux Libertine and Goudy Sans

Contents

Preface

There is general agreement among social scientists that institutions are crucial determinants of economic growth and human flourishing. The consensus is that they are more important than natural resources: a well-organized group of people can prosper in a harsh environment, while badly organized societies go astray even in rich and generous ones.

The bad news is that institutions are fragile: they cannot be taken for granted and require constant care. There are plenty of examples one could mention: the political institutions of my own country—one of the success stories of the past century—have been shuddering dangerously for almost two decades. The recent economic crisis has exposed the vulnerability of the institutions that govern global finance, and in some countries has put democracy to test, placing unelected technocrats in charge of reforms against the will of the people. These technocrats are often appointed by institutions like the European Union, which is itself in a state of crisis, according to many pundits and politicians.

These claims are familiar, and yet somewhat mysterious. What exactly is in a state of crisis? What are these things that must be reformed? Ultimately, what is an institution? If we do not know what institutions are, how can we possibly hope to improve their performance?

These are both philosophical and scientific questions. Philosophers have been asking "What-is-X" questions since the time of Socrates at least. Over the centuries however many of these questions have been taken over by science. If we want to know what is matter, or light, or life, we now ask physicists and biologists as well as philosophers. Similarly, questions about the nature of institutions cannot be answered satisfactorily without the help of science. So an important goal of this book is to offer a coherent picture of the fundamental architecture of modern societies, combining the insights of social scientists and philosophers who work on this topic.

Understanding Institutions aims at a broad audience: I have tried to write it in such a way that it could be understood by open-minded economists, political scientists, sociologists, anthropologists, and philosophers. I stress "open-minded" because the concepts and terminology that I use do not always fit those that are commonly employed in some of these disciplines. But it is inevitable, given the lack of unity that thwarts interdisciplinary research in the behavioral sciences. In fact one of my goals in writing *Understanding Institutions* was to introduce a unified theoretical framework that would facilitate conversation across disciplines in this important area of research. Another goal was to show that an adequate understanding of the nature of institutions helps resolve old conceptual and methodological problems in the philosophy of social science. While some of these problems simply disappear, others become more tractable once they are seen from the perspective of the unified theory.

The book owes a lot to numerous friends and colleagues whom I have had the luck to meet and to work with over the years. My first debt goes to my former colleagues at the University of Exeter. As a graduate student I found philosophical debates on the nature of society rather uninteresting, but I began to change my mind when Barry Barnes and Nigel Pleasants introduced me to the Wittgensteinean tradition in the philosophy of social science, and proved by way of example that my preconceptions were wrong. Supervising Mattia Gallotti's doctoral dissertation on collective intentionality was another crucial formative experience in the same period; I am grateful to Mattia for the numerous conversations that we have had and continue to have on these topics.

An important event for the genesis of this book was a seminar on rules and institutions that Frank Hindriks gave in Milan in the summer of 2011. In his doctoral dissertation, written a few years earlier, Frank had shown how to derive constitutive from regulative rules. Although I had read his dissertation back then, however, I had spectacularly failed to see the importance of this result. When Frank presented the same ideas in Milan, I realized that they could be used to build a unified theory of institutions based on the game-theoretic notion of correlated equilibrium. The unified theory has been outlined in two articles coauthored with Frank, and constitutes the bulk of the first part of the book. Although Frank and I do not agree on everything, it is fair to say that this book would have been a lot different without his contribution.

Over the years many other friends and colleagues have helped me. Cristina Bicchieri, Francesca De Vecchi, Natalie Gold, Conrad Heilmann, Muhammad Ali Khalidi, Arto Laitinen, Uskali Mäki, Chrys Mantzavinos, Luigi Mittone, Ivan Moscati, Fabienne Peter, Giacomo Sillari, Corrado Sinigaglia, Luca Tummolini, and many anonymous reviewers have provided useful comments on specific topics. I owe particular gratitude to a group of colleagues who have read the semifinal draft of the manuscript, have given constructive suggestions, and have spotted a number of mistakes that I had overlooked. They are, in alphabetical order, Matteo Bianchin, Frank Hindriks, Chiara Lisciandra, Olivier Morin, Michiru Nagatsu, David Teira, Enrico Terrone, Jack Vromen, and Jesus Zamora Bonilla. The published version is far superior to its predecessor drafts thanks to their comments. I'm aware that I have been unable to fully meet their expectations, but I hope they will forgive me: some of their demands were very demanding indeed.

Geoffrey Hodgson organized a wonderful symposium on one

of the papers that Frank and I wrote together, with commentaries by Masahiko Aoki, Ken Binmore, John Searle, Vernon Smith, Bob Sugden, and Geoffrey himself. I am very grateful to all of them for pushing us to correct and clarify our views on institutions. I am grateful also to the Ecole Normale Superieure for its hospitality in Paris during February 2014, and for the opportunity to give a series of seminars on the topics of the book. Daniel Andler, Mikael Cozic, Brian Hill, Elizabeth Pacherie, Cedric Paternotte, and some very clever graduate students offered useful comments back then and on subsequent occasions.

The main arguments of the book have been tested in several seminars and conferences. Since it would take too much space to mention them all, I thank the members of these audiences collectively. I must also thank Sarah Caro at Princeton University Press for believing in the project and steering it to completion. Special thanks are due to Andrea Branzi, who has generously given permission to reproduce a few drawings from his "strange little book," *Genetic Tales.* The images are reproduced by kind permission of Andrea Branzi and Alessi (©Alessi s.p.a., 1998).

The book is scattered with material borrowed from the following articles:

"Infallibilism and Human Kinds," *Philosophy of the Social Sciences* 40 (2010), pp. 244–64.

"The Normativity of Lewis Conventions," *Synthese* 190 (2013), pp. 3107–22.

"Reflexivity and Equilibria," *Journal of Economic Methodology* 20 (2013), pp. 397–405.

"On the Nature of Social Kinds," in *Perspectives on Social Ontology and Social Cognition*, edited by Mattia Gallotti and John Michael (Dordrecht: Springer, 2014), pp. 57–68.

"Institutions, Rules, and Equilibria: A Unified Theory," *Journal of Institutional Economics* 11 (2015), pp. 459–80 (with Frank Hindriks).

"Understanding Institutions: Replies to Aoki, Binmore, Hodgson, Searle, Smith, and Sugden," *Journal of Institutional Economics* 11 (2015), pp. 515–22 (with Frank Hindriks).

"A Unified Social Ontology," *Philosophical Quarterly* 65 (2015), pp. 177–201 (with Frank Hindriks).

"Philosophy of Social Science: Naturalism and Antinaturalism," in *The Oxford Handbook of Philosophy of Science*, edited by Paul Humphreys (Oxford: Oxford University Press, in press).

I thank Cambridge University Press, Oxford University Press, Routledge, Sage, and Springer for granting permission to reproduce this material. From 2009 to 2013 I was supported by a Rientro dei Cervelli scholarship of the Italian Ministry of University and Scientific Research (MIUR). Earlier work on conventions had been made possible by an ESRC grant (RES-000-22-1591) and by the generous support of the Computable and Experimental Economics Laboratory of the University of Trento. The Faculty of Philosophy at San Raffaele University provided much-appreciated hospitality during a delicate transition period in 2008–9. I am grateful to all the faculty members and especially to Michele di Francesco and Matteo Motterlini for making it happen.

Finally, I would like to dedicate this book to my colleagues in the Department of Economics, Management, and Quantitative Methods at the University of Milan, for their open-mindedness when they welcomed a philosopher of economics to their institution.

Analytical Table of Contents

Chapter 4 CORRELATION 44

In a correlated equilibrium each individual follows a rule of the form "if X, do Y," where X is an event external to the game. This approach is able to reconcile the rules and the equilibria accounts to the study of institutions: conventions are both behavioral regularities (equilibria), and regulative rules that guide and constrain the behavior of individuals.

Chapter 5 CONSTITUTION 57

According to John Searle, institutions are systems of constitutive rules of the form "X counts as Y in C." The distinction between regulative and constitutive rules however is untenable, because constitutive rules can be derived from regulative rules via the introduction of theoretical terms. In this chapter I show how it can be done, and incorporate Searle's account in the unified theory of institutions.

Chapter 6 NORMATIVITY 70

Institutional rules create rights and obligations, specifying actions that can or must be performed in certain circumstances. These deontic powers may be represented as costs that transform individual incentives in strategic games. This modeling strategy allows the extension of the unified theory to a wider class of games, including dilemmas of cooperation.

INTERLUDE

Chapter 7 MINDREADING 89

To achieve coordination, people must build concordant expectations about each other's behavior. Such expectations are sometimes based on the observation of public signals, but when public information is scarce we simply simulate the mental processes of the other individuals with whom we interact.

PART II APPLICATION

inductive inference, but they could be known directly and without error by the members of the relevant community.

Chapter 12 REALISM 163
The thesis of ontological dependence however is false: any social kind may exist independently of anyone holding a correct theory of that kind. There is no guarantee, for example, that people understand what money is, or that the things that people classify as money actually are money. The nature of an institution is determined by its function, not by what people think about it. As a consequence, we ought to be realists and fallibilists about social kinds.

Chapter 13 MEANING 177
The meaning of institutional terms is determined by the rules that people follow. But what if the rules are unsatisfactory? Can we change the rules without changing the nature of an institution? According to some philosophers we cannot. Thus, for example, we cannot use the term "marriage" to refer to the union of two individuals of the same sex. Sally Haslanger has tried to challenge this position introducing normative considerations for the identification of institutions.

Chapter 14 REFORM 194
Haslanger's approach unfortunately is inconsistent with realism. I argue that we can save realism and reformism by drawing a distinction between types and tokens. While institution tokens are particular solutions to coordination problems, institution types are identified by their function, or the kind of strategic problems that they solve. For example, same-sex unions are marriages because they fulfill some of the classic functions of marriage.

Introduction

Let's do an experiment: let's look around and describe what we see.

If I look to the right, I see my colleague Antonio reviewing a paper for a scientific journal. On the left, a map of China hangs on the wall. In front of me, past the door of my office, two students are walking in the corridor of the Department of Economics. Out of the window, I can see the second floor of the Faculty of Social Science of the University of Milan.

You and I, of course, are surrounded by different things. But if we compare our lists, they will have something in common: most of the things that we see are institutional entities. An "institutional entity" is an object with properties or characteristics that depend on the existence of an institution. Antonio, for example, is a colleague of mine because we are both employees of the same university, and the University of Milan is an institution.

When I drew my list I could have used a different language, describing all the things that I saw in noninstitutional terms. I could have focused on their physical, chemical, or biological properties, for example. But undoubtedly such descriptions would have been incomplete: most of the things that surround us are not just physical or biological entities. A description of the world that ignored institutional entities and properties would miss much of what constitutes our world.

But what are these entities then? Institutions are as mysterious as they are ubiquitous. We can mention some examples:

the church, democracy, the army, the public school system. But as soon as we try to theorize, we face many difficult questions. What are these things? What do they have in common? What are they made of, and how do they work?

In the chapters that follow I will try to answer these questions. Philosophers sometimes refer to this sort of inquiry using the term "ontology." This pompous name comes from the Greek words *on* and *logos*, meaning literally "theory or discourse about being." Ontology asks what there is in the world, and social ontology, in particular, studies what there is in the social world.

Social ontology has been characterized for decades by a Babel of different approaches. This has certainly made it a stimulating field of research, but at the same time has also created major problems of communication. Researchers have found it difficult to appreciate the value of alternative approaches, and often have decided to start from scratch, ignoring decades if not centuries of work on the same topics.

Problems of communication emerge frequently at the crossroads between social science and philosophy. Institutions are a major topic of research for political scientists, economists, sociologists, and anthropologists, so we should expect them to be the experts in this area. But philosophers have often found the theories of social science unsatisfactory. The author of one of the most influential philosophical books of the past twenty years, for example, has stated boldly that he cannot find anything helpful in the *entire* social science literature. The "tradition" is inadequate, "the classical theorists have the direction of analysis back to front." And this is true "not only of such foundational figures as Max Weber, Emil Durkheim, Georg Simmel, and Alfred Schutz, but of the whole Western tradition of discussing political and social institutions that goes back to Aristotle's *Politics*, if not earlier" (Searle 2005: 2).

The dissatisfaction is reciprocal: social scientists have struggled to appreciate what the contribution of philosophy might be. The book that I just mentioned for instance has been deemed "quite literally indifferent as sociology" (Osborne 1997: 98). And according to another reviewer, it shows "how big the hiatus between philosophy and the social sciences has become" (Knoblauch 1996: 1461).

It is an odd situation. It is certainly possible that different scholars have different concerns. Perhaps they approach the topic from different angles. But the general questions are clearly the same: What is an institution? What is the social world made of? How many kinds of social entities are there? So either the answers are substantially different, or they must be somehow compatible. In the first case, we must try to figure out which answers are right and which ones are wrong. In the second case, we must try to understand how seemingly different answers fit together; we must figure out whether they focus on different aspects of social reality, or whether they are expressing the same ideas using different vocabularies and theoretical frameworks.

In this book I propose a theory to unify the main traditions in the field of social ontology and explore the implications of this unification. In the course of the book I focus mostly on *human* sociality. This may seem a contentious decision, because humans are not the only social animals. From bees to hyenas, swallows, and chimps, many nonhuman animals live in groups and have interesting forms of social organization. But there is something special in humans: our societies are more complex and much more diverse than the societies of any other species. Throughout history, humans have experimented with many types of social organization, and there are other arrangements that are potentially feasible but have never been tried until now.

Consider the myriads of ways in which human beings have organized their family life. Anthropologists classify family structures as monogamous and polygamous, poligenous and polyandrous, exogamous and endogamous, matrilineal and patrilineal, matrilocal, patrilocal, bilocal, and neolocal, consanguinal, affinal, affiliative, and fictive (the list could be longer). And notice that the family is one of the social institutions that are most closely related to biological functions like mating and reproduction. Other institutions are even more autonomous from biological constraints, and have been shaped in thousands of different ways during the history of humanity.

So most of this book is devoted to understanding what human institutions are, how they work, why they differ, and what they can do for us. Since these questions can be tackled in different ways, it is important to pitch the inquiry at the right level. At the bottom of the scale, going from the particular to the general, we could study specific institutions like the British monarchy, the Catholic Church, or the Gandhi family. Following an established philosophical jargon, we shall call them *token* institutions. The main evidence that we are dealing with a token institution is that such entities have a history and a geographical location. They are situated in space and time.

Although historians and social scientists spend a lot of time studying token institutions, they theorize mostly about sets or classes of institutions. The point of theorizing is to generalize beyond specific cases, to explain by means of general models the functioning of more than one token institution. So the study of institutions will require that we shift one level up in the scale of generality. Terms like "monarchy," "political party," "bank," "firm," "union," "church," "cult," "family," "golf club" are used to refer to entire classes of institutions that share important properties. This is the level where one can hope to find models and categories of general theoretical interest.

There are many possible levels of analysis, which can be ordered hierarchically in terms of generality. "Church" refers to a kind of institution that is more general than "Protestant Church," which is in turn more general than "Reformed Church," and much more general than the "Dutch Reformed Church." A scholar of institutions may theorize at each of these levels, depending on her goals and interests. And in some cases she may even formulate general principles that hold for all institutions. This is the point where social science meets philosophy, in the field of social ontology. Social ontologists investigate the social world at the highest level of abstraction, and devise theories that are supposed to hold for all social institutions irrespective of their individual features.

Theorizing of course is easy, if unconstrained. The difficult task is to separate good from bad theories, those that are explanatory from those that are not. In order to do that, philosophers and scientists usually test their theories against particular cases. When a political scientist theorizes about types of democracy, for example, she tests her theories using token examples of historically existing democratic institutions. When doing social ontology we must proceed in a similar manner, except that the examples and counterexamples are usually picked from one level up: we test our theories against types of institutions.

There is obviously an enormous variety of types of institutions that can be used for this purpose. And the danger of picking examples to support one's favorite theory looms large. Fortunately, however, there is a set of cases that are considered paradigmatic and that must be accounted for by any theory that wants to be taken seriously. In the course of the book I will refer frequently to three types of institutions, either for illustrative purposes or to test specific theoretical hypotheses. My paradigmatic institutions are *marriage*, *private property*, and *money*. I will also occasionally refer to the rules of traffic, a simple institution that we are all familiar with, and that is analogous in many ways to more complex institutions like marriage, property, and money.

Each one of these institutions is a high-level type, which can be analyzed hierarchically into lower-level types and, low and behold, tokens. Marriage for example can be monogamous or polygamous, temporary or permanent, chosen or arranged, same-sex or different-sex, open or closed. And there are of course historical institutions with their peculiarities: there are Catholic, Muslim, Judaic, and Hindu marriages. There is my marriage with my wife, Caesar and Cleopatra's marriage, or the marriage of the gay couple who live next door. The interesting questions are, what do all these types and tokens have in common? What is *the institution of marriage* at the most general level of description? Answering these questions now would be premature. But as an appetizer, we can begin to notice that institutions are usually grouped by scientists according to their *functions*. The token institutions that anthropologists classify in the "marriage" category, for example, usually regulate activities aimed at procreation, the rearing of children, the care of the elderly, inheritance, and economic cooperation between the spouses. Similarly, money is defined by economists as whatever entity or type of entity is used as a store of value, medium of exchange, and unit of accounting ("money is what money does," as the saying goes).

The advantage of functional definitions is that they abstract away from the innumerable ways in which a goal may be achieved in different contexts. For this reason functions are used for classificatory purposes not only by social scientists, but also by biologists when they theorize about physiological traits. An eye, for example, is an organ that perceives and represents the environment through the detection of light. Eyes come in different guises, eyes can be classified according to different types, and each token eye (the eye of a wasp as opposed to the eye of a mammal) may exploit different light-detection mechanisms. Nevertheless, there are general theoretical principles that hold for eyes across the species. And similarly, there are interesting generalizations that apply to different marriages, regardless of the specific ways in which each token institution works.

The notion of function is strictly related to the idea of purpose or goal. So what is the purpose of institutions? As a first approximation, it seems that institutions facilitate coordination and cooperation. They help groups of individuals to do things that are better done together. Sometimes these collective activities are not particularly problematic, and coordination takes place easily. But at other times, the same goal can be achieved in different ways, each way implies a different division of labor, and it is not clear which is the best way to do it. In such cases, we shall say that there is a *problem of coordination.*

For example, it is easier to organize a dinner party if each host takes a specific role: you go shopping and I cook, I wash the dishes while you entertain the guests. For the party to be a success it is important that we all do our job. It would be a disaster if we both went shopping and no one cooked. But who is going to do what? Perhaps we both like cooking and we both hate shopping. At the same time, each one of us would be willing to do the shopping, if she knew that the other was doing the cooking. The problem is in part a problem of assurance, of being confident that the others are going to do their part in a complex collective task.

How can institutions build this assurance? Answering this question will take up the first part of the book. I will survey different views of institutions, analyze them critically, and explain how they relate to each other. I will begin by drawing a distinction between those theories that view institutions as rules, and those that view institutions as equilibria of strategic games. Then, I will argue that these two approaches are complementary, and that they can be unified within a single framework.

The *equilibria* approach spans across the divide between philosophy and social science. The seminal theory in this tradition was proposed by David Lewis in a justly celebrated book on *Convention* (1969), but over the past four decades several other philosophers and social scientists have proposed equilibrium-based accounts of social institutions. Theories within the equilibria approach view institutions as behavioral patterns that tend to persist because individuals have no incentive to deviate from the pattern unilaterally (unless everyone else does the same).

In spite of its explanatory achievements and its mathematical elegance, the equilibria approach has not been universally endorsed however. According to an equally popular alternative, institutions should rather be conceived as *rules* that guide the actions of individuals engaged in social interactions.

The rules account is close to our vernacular, prescientific understanding of institutions: intuitively, institutions regulate behavior, making certain actions appropriate or even mandatory in specified circumstances. The institution of private property, for example, regulates the use of resources by indicating who has access to them. The institution of money regulates the use of paper certificates in economic transactions. And the institution of marriage regulates the behavior of two or more individuals who pool their resources to raise kids, manage property, and help each other in many different ways.

But if institutions are rules, how do they influence behavior? Stating a rule is clearly insufficient to bring about an institution. To realize why, consider that there are plenty of *ineffective rules*: rules that are officially or formally in existence but that are nevertheless ignored by the majority of people. Traffic lights in Milan are regulation, in Rome they are a suggestion, and in Naples they are just decoration, as the saying goes. But since the rules are formally the same in Milan, Naples, and Rome, there must be something else going on. There must be some special ingredient that makes people follow the rules in some circumstances and ignore them in others.

The equilibria account of institutions tells us what the special ingredient is: effective institutions are backed up by *a system of incentives and expectations* that motivate people to follow the rules. An equilibrium in game theory is a profile of actions or strategies, one for each individual participating in a strategic interaction. Each action may be described by a simple sentence of the form "do X" or "do Y." The defining characteristic of an equilibrium—which distinguishes it from other profiles—is that each strategy must be a best response to the actions of the other players or, in other words, that no player can do better by changing her strategy unilaterally. If the others do their part in the equilibrium, no player has an incentive to deviate.

Since the actions of a strategic game can be formulated as rules, equilibrium-based and rules-based accounts of institutions are compatible. From the point of view of an external observer, an institution takes the form of a regularity that corresponds to the equilibrium of a coordination game. But each equilibrium strategy also takes the form of a rule that dictates each player what to do in the given circumstances. By combining the rules account with the equilibria account we obtain a unified theory that I call the *rules-in-equilibrium* approach to the study of institutions. Rules by themselves lack the power to influence behavior, but together with the right system of incentives and beliefs, they can influence the behavior of large groups of individuals. Institutions, in a nutshell, are rules that people are motivated to follow.

Institutional rules sometimes simply state that we must "do X" or "do Y." In many cases, however, they are *conditional* statements that prescribe different actions depending on the occurrence of certain events ("if X then do Y") . For example, the rules of traffic state that you must stop at the crossroads if the traffic light is red, proceed if it is green. Similarly, in many societies the actions of individuals are regulated according to their identities—there are rules of courtesy like "ladies first," as well as hierarchical rules like "give orders if you are the husband, follow them if you are the wife." Biological traits in such cases are used as signals that facilitate coordination, pretty much as traffic lights help us drive around smoothly. (If you are perplexed by this statement, let me clarify that these arrangements are not necessarily good equilibria: perhaps we would be better off if women gave orders and men obeyed; similarly, we could stop when the light is green and proceed when it is red.)

Traffic lights and biological traits are *correlation devices*, and the actions of people who use these signals constitute *correlated equilibria.* Correlation devices multiply the number of ways in which we can try to coordinate. Suppose, to use my trite example, that you and I want to organize a dinner party. To simplify, let us suppose that we do not have strong preferences regarding the division of labor. To make sure that we coordinate, I text you a message: "I shop and you cook." The main purpose of this signal is to create the expectation that I will go shopping. Because if you believe that I will go shopping, then you will do the cooking, and the party will be a success. But of course this is just one of many possible signals that we could have used to coordinate. Had I told you "I cook and you shop," the opposite equilibrium would have been implemented. So language is a tremendously versatile device to create institutions, by sending signals that people use to converge on new equilibria. Humans are special in the animal kingdom in large part because they have language, and because they can use it to create a wide range of different social arrangements.

This point has not passed unnoticed of course. The most original and systematic attempt to place language at center stage in social ontology is the theory of constitutive rules proposed by John Searle. Although this theory is a variant of the rule-based account of institutions, it attempts to explicate institutions using a very different kind of rule that, instead of merely regulating behavior, creates the possibility of new types of behavior. Constitutive rules according to Searle are statements of the form "X counts as Y in C," where Y denotes an institutional entity or fact or property, X is a preinstitutional entity, and C is a set of circumstances or conditions of instantiation. In the case of money for example a constitutive rule is: "Bills issued by the Bureau of Engraving and Printing (X) count as money (Y) in the United States (C)" (Searle 1995: 28).

Searle contrasts constitutive rules to regulative rules that have as their syntax "do X," or "if X do Y." The actions or strategies that appear in game-theoretic accounts of institutions, as we have seen, have precisely this form, so Searle's distinction suggests that there is a deep hiatus between his own approach and the accounts of institutions found in the social science literature. But if this were true, then the attempt to unify different approaches to social ontology would fail: not all institutions would be systems of (regulative) rules in equilibrium.

There are good reasons, however, to believe that Searle's distinction between regulative and constitutive rules does not hold. Using an argument originally devised by Frank Hindriks, I will show that constitutive rules have a much more limited role than the one envisaged by Searle: they are term-introducing principles that state the conditions of application of the theoretical terms that we use to label institutions. They are, first and foremost, naming devices for regulative rules.

The constitutive rule of money, for example, specifies the conditions that have to be satisfied for something to be money (it must be a paper bill issued by the Bureau of Engraving and Printing), and implicitly specifies what to do with paper certificates of that kind (use them to trade commodities, save them for future purchases, etc.). Hindriks's view that regulative rules can be transformed into constitutive rules via the introduction of theoretical terms highlights the fact that constitutive rules do not add anything that cannot be expressed by means of simple regulative rules. In principle they could even be eliminated from our theoretical vocabulary, without causing any substantial ontological loss. The constitutive rule of money for example can be translated in a regulative rule such as: "if a bill has been issued by the Bureau of Engraving and Printing, then use it to purchase commodities or save it for the future," and so forth.

The unified theory thus helps attain ontological parsimony and at the same time offers an explanation of the pragmatic function of institutional terms (why they are useful and how they help us coordinate). Having accomplished that, the remaining part of the book will be devoted to articulate the theory in more detail, and to explore its philosophical implications. In particular, I will focus on the implications of the unified theory for the explanatory and predictive ambitions of social science.

For well over a century social scientists have been discussing the methodological foundations of their discipline. On the one hand, methodological "monists" have been arguing that the social sciences must follow the same approach as the natural sciences. On the other hand, methodological "pluralists" have argued that the very nature of social reality makes it impossible for social scientists to attain the same explanatory and predictive success of the natural sciences. Social scientists should adopt a different approach and give up the traditional goals of naturalistic scientific inquiry.

What ontological differences may license this kind of skepticism? A classic cause of concern has been the *mind-dependence* of social reality. The idea is that social entities differ from natural entities in that the former, but not the latter, depend essentially on our representations. The nature of a dollar bill, the fact that it is money, for example, depends on a collective belief or recognition that it is money—that it can be used to buy certain commodities and services. (Otherwise, it would be just a piece of paper with a picture of George Washington printed on it.) In contrast, a molecule of water is water regardless of what anybody believes about it. It does not have to be represented as water, in order to be what it is.

The thesis of mind-dependence has been used by many theorists to challenge the scientific ambitions of social science. The challenge can take different forms, however, depending on how the concept of dependence is interpreted. So part of the book will be devoted to distinguish between different versions of the dependence thesis. In particular, it will be useful to distinguish between *causal* and *noncausal* dependence on representations.

I will argue that the thesis of causal dependence is true, but that its philosophical consequences have been exaggerated. Mind-dependence, in particular, when it is interpreted causally does not constitute a threat to the scientific ambitions of social science. The thesis of noncausal dependence, in contrast, is just false. I will try to give a precise formulation of the thesis using the notion of ontological dependence, and show that it is inconsistent with the functionalist understanding of institutions that is common in the social sciences. This, as we shall see, implies that we can entertain a realist and fallibilist attitude toward the entities studied by social scientists, just as we do with the entities studied by natural scientists.

The final two chapters are devoted to an issue that is currently hotly debated in many countries, concerning the design and identity of one of our most important institutions. The issue is whether to reform the institution of marriage so as to make it possible for partners of the same sex to get married. As we shall see, traditionalists have claimed that the institution of marriage is intrinsically or necessarily limited to heterosexual couples, and that the inclusion of same-sex couples would turn it into a different institution. The claim has often been backed up by sophisticated semantic arguments, and philosophers have been engaged in the battle on both sides of the field.

My own view is that it is perfectly legitimate to use the term "marriage" to refer to the contracts that regulate the relationships between individuals of the same sex. However, the debate on marriage highlights an interesting problem: it suggests that it is difficult to be simultaneously a realist and a reformist about institutions. Some philosophers have argued that the identity of institutions depends not on the rules that people actually follow, but on those that they *should* follow—that is, on the normative targets that we set for ourselves as a community. This "ameliorative" approach (a term used by Sally Haslanger) however is incompatible with realism. So I will propose a different solution based on the unified theory, to save both the realist principle that institutions do not depend noncausally on our intentions, and the reformist intuition that the rules of the game can be redesigned without changing the identity of an institution.

This is, more or less, the content of this book. Although I do not expect anyone to be persuaded by a short summary, I hope the appetizer will make you want to read more. There are gaps in the argument, and I will try to fill them in the chapters that follow. But even if I fail to convince, I hope the book will help philosophers and scientists appreciate how different projects in social ontology are related to one another. I hope that it will foster communication across research programs, and collaboration between scholars who adopt different approaches. Social ontology has been disunified for too long, and it is time that we put it together again.

REFERENCES AND FURTHER READINGS

Social ontology is a lively field, and the literature as a consequence is quite large. The volumes edited by Schmitt (2003), Mantzavinos (2009), and Gallotti and Michael (2014) include essays by many influential philosophers and are a good point of entry in contemporary debates. The diversity of human institutions and their relative independence from biological constraints are central topics in contemporary social science and have generated heated debates. For a view that emphasizes biological constraints, influenced by evolutionary psychology, see Boyer and Petersen (2012). On functionalism in social science and biology, I tend to follow Pettit (1996). Since most of the topics mentioned in this introduction are analyzed in more detail in the rest of the book, I refer the reader to the literature reviews at the end of the subsequent chapters.

I

UNIFICATION

CHAPTER 1
RULES

> Institutions are the rules of the game in a society or, more formally, the humanly devised constraints that shape human interactions. . . . They are a guide to human interaction, so that when we wish to greet friends on the street, drive an automobile, buy oranges, borrow money, form a business, bury our dead, or whatever, we know (or can learn easily) how to perform these tasks.
>
> —North (1990: 3–4)

The most famous and cited definition of social institution appears right at the outset of Douglass North's monograph on *Institutions, Institutional Change and Economic Performance.* North is primarily an economic historian, known for his work on trade and growth in early modern Europe. His writings however have influenced many scholars outside his field of specialization, and for this reason in 1993 he was awarded the Nobel Prize in economics. The conception of institutions as rules is not North's invention, to be sure. Similar definitions can be found in the writings of prominent social scientists of the twentieth century, like Max Weber, Talcott Parsons, and Friedrich Hayek. In philosophy, rule-based theories are equally popular—an original version proposed by John Searle will be discussed later in the book.

The conception of institutions as rules is intuitive, and fits with our pretheoretical understanding of many paradigmatic institutions. Consider marriage for example: the state of being married is associated with several rights and obligations. In most Western countries both husband and wife are responsible for procuring the material resources that are necessary to support the family. They are responsible for their kids' well-being and education. Moreover, they share a mutual obligation to be faithful and to help each other in case of need.

At the level of token institutions these general principles are translated into more specific behavioral rules that govern the division of labor of the spouses in their everyday tasks. Some rules regulate chores ("I cook, you wash dishes"), others regulate child care ("I change nappies, you feed the baby"). Some rules concern finance, others concern sexual behavior, and so on and so forth.

The reason why such rules exist is fairly obvious: they help husband and wife achieve goals that would be difficult to attain if they acted independently. If they both devote a lot of time to cooking, but nobody feeds the children, the kids are going to starve. If they both look after the children but no one goes to work, there will be nothing to cook tomorrow. As an analogy, consider a team of basketball players: if they follow their coach's assignments (he runs, you pass the ball; she defends, you attack) a group of players can hope to win matches and trophies. Without rules, in contrast, they will probably lose every single game. Another example that recurs in discussions of institutions is with the rules of traffic: institutions regulate individual actions in such a way that everyone can benefit from orderly behavior, just like complying with traffic rules is generally beneficial to drivers. Unpleasant events—accidents, jams, disputes—are avoided, or at least their frequency is significantly reduced if we all follow the rules.

Two points must be clarified before we move on. First, the idea that institutions are beneficial is dubious and perhaps even meaningless unless we specify clearly the contrast case: they are beneficial compared to what? Second, that institutions are generally beneficial does not mean that they benefit all individuals in the same manner. It is easy to find examples of unequal or unfair institutions: in traditional marriage arrangements, for instance, women are often burdened with more obligations and fewer rights than men. Similarly, servants are definitely worse off than their masters under the institution of slavery.

The latter example is particularly controversial: how can slavery be "beneficial," given that the institution itself is the main cause of servants' misfortune? The answer is that we are not comparing the welfare of slaves under this terrible institution with the welfare they could enjoy in a more humane institutional arrangement. The right comparison is with the welfare they would enjoy in a noninstitutional arrangement. Historically, slavery has tended to arise whenever the asymmetry of power between two social groups has been so large that one of them could easily exterminate the other. The enslavement of Native Americans in the sixteenth century, for example, was the consequence of the superior military technology, organization, and resistance to diseases of the European conquerors. Slavery thus was "beneficial" to the slaves only in the grim sense that the noninstitutional alternative for Native Americans would have been genocide. So the point is merely that institutions improve people's lives compared to a situation without institutions, in which they behave independently without the guidance of rules. Institutions, in a nutshell, are better than chaos.

This is entirely compatible with the fact that many people might be better off under an alternative institutional arrangement. There is rarely a unique way of regulating our lives. In a basketball team, I might be the shooting guard and you might be the center, or the other way around. In a family, the husband may stay at home and the wife may go to work, or vice versa. Each particular institution—who does what, or who fills which role—allocates burdens in different ways, and consequently some people may prefer one type of institution to another. Occasionally, it may happen that we would all be better off under an alternative arrangement. People sometimes get stuck with bad institutions because they cannot decide to change the rules, or because they are not sure that new rules will be followed, or simply because they cannot see that a better institution is available.

One final remark on "beneficial" is in order before I proceed. The term refers only to the group of people whose behavior is regulated by the institution. Since institutions are often group-specific and exclude many people, it is possible that the benefits they confer to the members of one group are offset by the negative effects they have on the members of another group (the outsiders). A typical case is the Mafia, an institution governed by rules of secrecy, cooperation, obedience that benefit the mobsters but harm their victims. But even legal institutions like the army may have positive consequences for some individuals (the soldiers, the people they protect) as well as extremely negative consequences for others (the enemies and civilians who are killed during a war, for example).

Having said that, it is hard to deny that *in general* the capacity to regulate collective behavior is a tremendous asset for our species. The spectacular demographic growth of *Homo sapiens* and its rise to supremacy on Earth are due in large part to its social skills and flexibility of organization. Institutional economists like North have studied in particular the role played by institutions in facilitating economic growth. The idea—confirmed by a wealth of empirical studies—is that rules can help overcome obstacles that limit production, trade, and more generally hinder the welfare of a society. (Economists use the technical term "transaction cost" to refer to these impediments.) New rules may be created by an influential group, for example an enlightened ruler or government. However they may also emerge and evolve autonomously, without anyone in particular planning or foreseeing their effects. If they are successful, institutions are often spontaneously copied and disseminated across different social groups. But again, this is by no means guaranteed: clever ideas sometimes do remain unrecognized.

For historical and cultural reasons, a lot of research carried out in the past century has tended to emphasize the spontaneous emergence and diffusion of institutions. This was partly a reaction against an older approach to social policy that emphasized government intervention and central planning. Scholars interested in the spontaneous evolution of institutions draw a distinction between *formal* and *informal* institutional rules. "Formal" here means stated explicitly, codified in a set of laws, principles, rights that are publicly available and known or at least knowable by the relevant members of society. Such rules may be transmitted orally, but in complex societies they are usually preserved in written form. Informal rules, in contrast, are not explicitly codified and become manifest mostly through the behavior of individuals.

Friendship, for example, is governed entirely by informal rules: although there is no formal rule stating that you should not date your friend's boyfriend, it is generally agreed that it is not to be done and transgressions may cost you dearly. A complex institution like marriage in contrast is constituted both by formal and by informal rules. There is an important asymmetry between informal and formal institutions: while "purely" informal institutions are quite common, it is difficult to find examples of institutions that consist exclusively of formal rules. Even written legal codes rely heavily on informal practices for their interpretation and implementation. The fact that a certain rule is formally included in the body of principles that constitute the laws of a country actually has little significance in itself. Many laws are never followed and their transgressions are never punished, in spite of the fact that no one has bothered to abrogate them formally.

In May 2010 ten French ministers proposed to repeal a law

forbidding women to wear trousers. The law had been in place since 1799, although hardly anyone had noticed for a long time. When it was finally recognized as invalid, in 2012, the official act of the French Parliament had mere symbolic significance. Rules like the French ban on trousers are *ineffective.* The distinction between effective and ineffective rules is very significant from a theoretical point of view, because it is strictly related to a deep problem of the rule-based conception that will keep us occupied for the rest of this chapter.

Effective rules are important for policy making because institutions are causal factors that can be manipulated to achieve certain goals, by changing people's behavior. For example, the introduction of an institution that provides insurance to farmers may change their business practices and improve the efficiency of their firms. The introduction of a rule of hygiene may reduce the incidence of stillbirths and improve the fertility of young women, and so on. The rule-based approach, by itself, however does not explain why people comply with the rules. Why are *some* rules followed and not others? This is not just a philosophical question. It is an extremely important practical issue, because if we do not know the answer we risk designing institutions (rules) that fail because people do not comply with them.

Another way to put it is this: rules are linguistic statements, but to state a rule—to say "do this," or "do that"—is not enough to create an institution. The case of the French law is an obvious example. In that case, the law was simply forgotten, but the problem is deeper: some rules are not implemented even though they are widely known. In several North American states for example the speed limit on the motorway is officially sixty-five miles per hour. However most cars drive between sixty-five and seventy-five. So clearly the formal rule is not effective—the *real,* informal rule sets the speed limit somewhere around seventy-five. But to say that sixty-five is not the "real" rule leaves many important questions unanswered: What distinguishes "real" from "nominal" rules? What is the difference between the sixty-five- and seventy-five-miles-per-hour rules? Why do people comply with the latter but not with the former?

A plausible explanation may go like this: although the formal rule sets the limit at sixty-five, there are advantages in being slack in its enforcement. A driver speeding at sixty-five for example may find herself in a situation where she should swiftly accelerate to avoid an accident. If she believes that small breaches of the rule are going to be fined, the driver might hesitate, with catastrophic consequences. Given that traffic rules are meant to reduce the number of accidents and improve safety, it is wise to leave a little room for maneuver around the official speed limit.

Second, there is a problem of measurement: measurement instruments are imprecise. This is true of both the instruments that are available to the drivers and those used by the police. Fining cars that speed at sixty-six miles per hour would generate a lot of litigation, appeals, discontent, accusations of unfairness. So it may be wise for the police to sanction only major violations of the traffic rules. In practice the police may decide to implement a strategy like this: fine every car speeding at seventy-five miles per hour or more; fine some cars speeding at seventy to seventy-five; fine no car speeding at sixty-five to seventy. This strategy would work reasonably well and ensure that most people drive right around seventy. But those who are caught speeding at seventy-five or more cannot complain if they get a ticket: they were clearly exceeding the official speed limit.

The above explanation depicts an effective rule as a stable state. Drivers have an incentive not to exceed seventy-five miles per hour; the police have an incentive to tolerate those who do not exceed that limit (because they do not want to waste too much time litigating). If a naïve observer were to look at the traffic flowing down the highway, she would conclude that the effective speed limit is roughly seventy-five: everybody's behavior confirms the expectation that one should not exceed that limit. The system is in equilibrium.

Three points are worth keeping in mind: First, simply to say that institutions are rules does not explain why people follow some rules but not others. Second, by trying to explain why the sixty-five-miles-per-hour rule is merely nominal, while seventy-five rule is effective, we are induced to analyze the factors (especially the incentives) that promote compliance. And finally, such factors take a special configuration—a kind of equilibrium state.

The preceding line of argument suggests that institutions must be *special* kinds of rules. They are better conceived of as rules that people have an incentive to follow. The concept of incentive and the concept of equilibrium are strictly related. Intuitively, a system is in equilibrium if the forces that determine its current state contribute to perpetuate it indefinitely. Take a book and put it on your desk, for example. The force of gravity pulls it toward the center of the Earth, but its fall is impeded by the force of cohesion of the molecules that constitute the surface of the desk. Unless some other force intervenes, this state of affairs will continue to hold indefinitely—the system is in equilibrium.

In the case of institutions, of course, the forces at play are different. Human behavior is affected by a large number of causal factors, and it would be foolish to try to give a full list. When social scientists are dealing with very general issues such as the nature of institutions, they use modeling tools that do not make a precommitment to any specific mechanism. They simply assume, for example, that the behavior of individuals is governed by "incentives." An incentive is a property of a state of affairs that motivates people to action. Incentives need not be material goods like food, sex, or shelter, and people need not be motivated by purely economic interests. It is perfectly legitimate to assume that people have different ultimate goals in life. If one's goal is to glorify the Almighty God, for example, it may be in one's interest to spend a lot of money to build a magnificent cathedral. If one's goal is to raise healthy and happy children, one may have an incentive to invest in education. Incentives in this sense do not necessarily depend on self-interest narrowly conceived.

A social system is in equilibrium when the incentives of the relevant actors contribute to keep it in its current state. A stable social state is not necessarily motionless like a book lying on a desk. It may involve frantic activity, and the actions of one individual may be quite different from those of another. But a social system in equilibrium is likely to be characterized by *regular patterns of behavior*: people will tend to do (roughly) the same things in the same circumstances. Thus, for example, a Martian who has just landed on Earth may notice that North American drivers tend not to exceed the seventy-five-miles-per-hour limit, with remarkable regularity.

Andy Schotter—a prominent game theorist and experimental economist—defines institutions as "regularities in behaviour which are agreed to by all members of a society" (1981: 9). Such regularities "can be best described as noncooperative equilibria" of strategic games (1981: 24), because out-of-equilibrium actions are unstable and are unlikely to be repeated in the course of many interactions. This definition summarizes the main elements of the equilibria approach to the study of institutions. Like its main rival (the rules account) the equilibria approach is more like a tradition or research program than a single theory. And like its main rival, it cuts across the divide between science and philosophy.

Historically, the equilibria tradition stems from the study of *coordination* problems. Coordination problems are ubiquitous in social life, and are an important breeding ground for institutions. The rules of traffic are attempts to solve problems of this kind: we could set the speed limit at sixty, sixty-five, seventy, or any other figure that offers a reasonable compromise between speed and safety. Exactly which one we choose is not very important, it is a matter of convention. Similarly, we could all drive on the left- or on the right-hand side of the road. Either solution is fine, provided we all do the same thing. In other cases it is important that we do *not* all do the same thing: in politics and war, it is often important that someone leads and the others follow for example. If we all follow or we all lead we are going to be in trouble. In a marriage, it is important to divide labor—I cook and you wash the dishes, or you cook and I wash. What these situations have in common is that we all have an interest to coordinate, and that there is more than one way to do it.

How are coordination problems solved, in practice? One obvious way is by stating a rule. When they arrive in Dover, for example, several signs remind Continental Europeans that the rule is "drive on the left." It is important to stress, however, that people do not drive on the left in Britain just because of that rule. To realize that this is so, it is sufficient to engage in a little thought experiment. Suppose you woke up one morning and discovered that everyone is driving on the right in Great Britain. What would you do? Even if the rule officially has not changed, clearly you would also drive on the right. You would do it because it would be in your interest. People do not follow the rules of traffic because a piece of legislation says so, but because they do not want to crash.

If you know what the other individuals are going to do, then coordination is not a big problem. And in many cases, you can simply see what they do and then decide accordingly. Imagine there are two free tables at a restaurant; you see that another customer is looking for a place to sit. If you are indifferent between the tables, you can just wait and see which one the customer is going to choose, before you make your move.

But what if you cannot observe others, for example because you have to make your decisions independently and simultaneously? Having a rule then may help coordination. Suppose there are only two chairs left in a crowded conference room, but one of them offers a better view of the stage. Two members of the audience approach them simultaneously. Perhaps they will simply move as fast as they can and scramble for a seat. But this may cause embarrassment and bad feelings. Now suppose the two individuals are a gentleman and a lady, or a young and an old person, or one of them is physically impaired. Then a simple informal rule of etiquette ("ladies first") may solve the problem of coordination.

For this to work everyone must be aware of the rule, and must believe that the others will follow it. Only if this is the case will it be in our interest to comply. But how do we know that the others will follow the rule? Clearly the others are facing exactly the same problem: it is in their interest to comply, provided that *we* comply. But since we have not decided what to do yet, how can they expect us to comply? They cannot solve their problem until we have solved our problem, and we cannot solve ours until they have solved theirs!

The problem of coordination looks puzzling when it is seen through the eyes of a logician. But real people are rarely bothered by logical paradoxes. Unlike Buridan's ass, who starved because it could not make a decision between two identical stacks of hay, real people at some point just go one way or the other. When pure logic fails, some extralogical mechanism breaks the tie.

One of the striking aspects of real-life coordination problems is that we rarely pause to think about them. In most cases the solution looks obvious, indeed so obvious that it does not look like a problem at all. If the sign says "Keep Left," then it is obvious that the thing to do is to keep left, and that the others will do their part. This is the most natural solution, the one that first comes to mind and that "stands out" from the crowd of all the possible solutions. So perhaps the main function of the sign (and the rule) is to make one solution appear obvious. Lacking a better reason, people will just opt for the solution that first comes to mind.

A solution that "stands out" from the crowd is a *focal point*, in technical jargon. This expression was coined by Thomas Schelling in a pioneering book, *The Strategy of Conflict* (1960), that laid the foundations for much contemporary work on institutions. Schelling noticed that individual decisions are facilitated by the existence of focal points. A focal point is often *salient* for reasons that have nothing to do with the structure of incentives of the game, but depend on features of the environment that are irrelevant from a purely logical point of view. Schelling pointed out that people sometimes make better decisions if they do not engage in complex logical reasoning—just as we do when we instinctively follow the sign.

There are many possible sources of salience. In many cases the focal point is determined by our cognitive processes. Consider Figure 1.1, for example. Suppose that two players have the opportunity to choose a star among those that appear in the picture, and if they choose the same star they will earn some money (ten euros, say). Which one would you choose? The answer is obvious because one star is immediately salient. From a purely logical point of view, the fact that it is roughly in the middle of the picture should be entirely irrelevant. Still, that star looks somewhat different—it "stands out" from the crowd. Many people choose the star in the middle and by so doing they raise the probability of coordinating, because it is the most frequently chosen star.

Not all focal points exploit perceptual salience though. Another important source of salience is *culture*. Let us consider a slightly different coordination problem, with four stars perfectly aligned in a row (Figure 1.2). Faced with this problem, the majority of people choose the first star on the left. This seems to be the obvious solution for reasons that have little to do with perception. Or, to be more precise, most of us first look at the star on the left. But we do not do it because it is visually striking, or "different." We do it because we have been taught to read from left to right. So we choose the star on the left because

FIGURE 1.1. A case of perceptual salience.

it is the first one, and it is the first one because in our culture we read and count from left to right. The solution is salient for cultural reasons.

This shifts the problem one step further though. What is "culture," in fact, if not a massive system of rules that govern our social interactions? And the rules surely are conventional, much of the time. Leonardo da Vinci wrote in Italian from right to left, as many contemporary people do when they write in Arabic, Hebrew, and Urdu. Whichever convention we use, we do it to facilitate coordination with others. So the reason why I occupy the right lane when I drive in my country is simply that most Italian drivers have been driving on the right up until now. Since we are all facing this problem every single day (which side of the road should I choose?) I cannot rule out that, for some reason, the other drivers will decide to change their behavior at some point. And it is not science fiction: Swedish drivers changed their convention overnight—from left to right—on September 3, 1967, between four fifty and five o'clock in the morning. (For ten minutes everything stood still, apparently.) But the Swedish government affixed a hundred thirty thousand signs to make sure that everybody knew, and that everyone was confident that everybody knew, that a major change of convention was going to take place.

Figure 1.2. A case of cultural salience.

It takes a massive effort to convince people that they should abandon an engrained custom, in fact. In normal circumstances the hypothesis that we could all switch from right to left or vice versa is totally bizarre. Thus among the "irrelevant" details that can make a strategy salient *history* plays an important role. As Schelling pointed out "precedence seems to exercise an influence that greatly exceeds its logical importance or legal force," and "there is . . . a strong attraction to the status quo ante" (1960: 67–68).

This simple insight proved to be tremendously fertile. Schelling's lectures at Harvard were attended in the 1960s by a young philosopher called David Lewis, who was looking for ideas to use in his doctoral dissertation. Lewis was particularly interested in the conventional aspects of language, and realized that Schelling's focal points provided a neat explanation of how linguistic conventions could emerge and persist in the course of repeated interactions. In his dissertation—which later became his classic book on *Convention* (1969)—Lewis pointed out that an outcome can become salient simply in virtue of the fact that it has occurred very frequently in the past. The focal point is determined by history.

So we drive on the right because everybody has driven on the right up until now; we interpret +2 as "add two" because everybody has done so until now. Ditto for wearing black at funerals, or using the word "shark" (rather than "whale," "stark," "skkrk," or whatever) to denote a dangerous fish that lives off the coast of Australia. When a behavioral regularity perpetuates itself in this manner, we will say that a convention has emerged in a given population.

Lewis says that there is no logical reason to choose the salient equilibrium, and no logical reason why precedence should be a source of salience. It is just a fact, based on our brute disposition to notice patterns, to extrapolate from the past, and to attribute the same extrapolating tendency to other human beings. We are built like that, and we should better take this fact into consideration when we do science or philosophy: "we tend to follow precedent, given no particular reason to do anything else" (Lewis 1969: 39).

But a disposition to follow precedence is not merely an *internal* propensity of each individual (biologically inherited or acquired by training). There is also a normative component in rule following: we tend to blame people who do not follow precedent or do not choose the obvious solution to a coordination problem. We feel that deviants do not just do something that is statistically anomalous, or rare. We also have the impression that they do something *wrong*.

Lewis sketches a simple story about the normativity that we attribute to conventions. A convention is normative in two senses: breaching it would damage the deviant, but also other members of the community. So there are at least two good reasons to conform.

> For we do presume, other things being equal, that one ought to do what answers to his own preferences. And we do presume, other things being equal, that one ought to do what answers to others' preferences, especially when they may reasonably expect one to do so. For any action conforming to a convention, then, we would recognize these two (probable and presumptive) reasons why it ought to be done. We would not, as far as we can tell, recognize any similarly general reasons why it ought not to be done. This is what I mean by calling conventions a species of norms. (Lewis 1969: 98)

The first "ought" is a norm of instrumental rationality (do what is best for you, in the given conditions). The second one is a social norm that prescribes not to harm others unless there is a good reason to do so. These two norms jointly suffice to consider certain actions as "right" and others as "wrong": the "right" ones are those that are expected by the community.

I will return again to the issue of normativity, because it plays an important role in the analysis of institutions. But for now, let me conclude this discussion of rule-based accounts of institutions. We have seen that the application of rules is determined in part—but essentially and inevitably—by the incentives of the community members. I started from the problem of distinguishing effective from ineffective rules, or how to explain why certain rules are actually followed while others are not. Looking at the case of speed limits, I argued that an effective institution is an equilibrium state where all the relevant individuals have an incentive not to deviate from a certain pattern. So any rule-based account of institutions is bound to be incomplete. Minimally, it should be complemented by an account of how and why some rules are followed and others are not. The examples discussed so far point in a certain direction, suggesting that institutions should be represented as equilibrium states. To articulate this idea in more detail it will be necessary to use some concepts borrowed from the theory of strategic games.

REFERENCES AND FURTHER READINGS

The idea of institutions as "rules of the game" (*Spielregeln*) is already in Weber (1910: 459). Accounts of institutions as rules can be found in the writings of Parsons (1935), Knight (1992), Mantzavinos (2001), Hodgson (2006), Miller (2010), and many other social theorists. The example of speed limits is borrowed from Greif and Kingston (2011). There is a striking analogy between the problem of ineffective rules and the famous problem of "meaning finitism" discussed by Wittgenstein (1953). Following Wittgenstein, Kripke (1982) has argued that the meaning of a term must ultimately be determined by a system of social incentives, a solution that is similar to Lewis's equilibrium account of conventions. See also Bloor (1997) and especially Sillari (2012), who defends this parallelism in detail. The books by Schelling (1960) and Lewis (1969) are extremely clear and accessible, so it is worth reading the original texts. Although they both continued to give important contributions to social science and philosophy, they did not write any other major book on these topics. They did however start an extremely influential research program that has produced a large literature. A favorite of mine in this tradition is Robert Sugden's *The Economics of Rights, Co-operation and Welfare* (1986), a monograph that I will cite again in subsequent chapters. Sugden (1998b) includes an excellent discussion of the relation between inductive propensities and the emergence of conventions. In Guala and Mittone (2010) and Guala (2013a) I present some experimental data on the normative power of conventions, a topic that has been discussed also by Gilbert (1989) and Tummolini et al. (2013). Although rules do not play a central role in their theory, Smit, Buekens, and du Plessis (2014) articulate an account of institutions as systems of incentivized actions that is in many ways similar to the one that I defend in this book.

CHAPTER 2
GAMES

The architecture of the equilibria approach to the study of institutions is based on game theory, and since game-theoretic concepts will be used again in the course of the book, it is a good idea to familiarize ourselves with them at the outset. A "game" is a situation of interactive decision making. In a typical game a set of players have the opportunity to choose an action from a menu that includes various options. The consequences of the action however are not entirely under the control of any individual decision maker: which outcome will occur depends partly on what the other players do.

What is the advantage of modeling social interactions as games? According to Thomas Schelling, "the most helpful invention of game theory for the social sciences is the payoff matrix" (2010: 29). A matrix is simply an array of symbols or numbers arranged in rows and columns. By modeling games as payoff matrices, we can represent various factors that play an important role in social behavior: actions, outcomes, agents, and incentives. As an example, take the matrix in Figure 2.1. The individual rows and columns, marked by capital letters, represent the *actions* or *strategies* of each individual player. By convention, Player 1 is Row and Player 2 is Column. In this case the players choose from a menu of two actions, called "Left" and "Right." The cells with numbers represent the *outcomes* or consequences of the actions.

	L	R
L	1, 1	0, 2
R	1, 1	3, 3

FIGURE 2.1. A simple two-by-two game.

In the course of this book I will make extensive use of two-by-two matrices. A two-by-two game involves two players with two possible actions for each player. Of course not all social interactions are as simple as that, but two-by-two games have the advantage of being easy to analyze and to represent visually. Moreover many claims derived from the analysis of these simple situations can be generalized to more complex settings.

Matrices are typically used to represent games with imperfect information. This means that when they make their decisions, the players do not know all the factors that will determine the outcome. In particular, they do not know the actions chosen by the other players. These situations are often called "simultaneous games," although the label is slightly misleading: what really matters is not that the players make their decisions at the same time, but that each player cannot observe the moves of the other player when she makes her decision. Clearly beliefs are very important in such games, because the actions of the players depend on their expectations about the strategies of the other players.

In a strategic game each outcome is represented by a set of numbers or *payoffs*: the first number is the payoff of Player 1 (Row), the second one of Player 2 (Column). For example, in Figure 2.1 the profile of actions LL yields one unit of payoff for Player 1 and one unit of payoff for Player 2 (1, 1). The standard view in game theory is that payoffs represent *preference* rankings, expressed in units of *utility*. For example, since $3 > 1 > 0$, Player 1's (Row) most preferred outcome is RR, followed by RL and LL; the least preferred outcome is LR. Notice that preferences are properties of individual players—subjective evaluations or individual propensities to behave in a certain way. But individuals usually have preferences about objective features of the external world, such as food, shelter, the number of hours spent idling on the beach, and so forth. Since preferences may not be easily observable, it is sometimes useful to interpret payoffs directly as measures of objective properties of the world or objective outcomes. In general, we shall focus on things that people care about, and unless otherwise specified we will assume that people prefer to have more rather than less of these things.

A matrix provides the resources to represent a situation of strategic interaction, but does not tell us how the players are going to behave. Even when we know their preferences, it is often difficult to determine what people will or should do. The search for a plausible *solution concept* (a concept that identifies the combination of actions that will be chosen) has been from the beginning one of the central concerns of game theorists. However, to quote Schelling again, it has turned out that "game theory is great at exploring situations, less able to provide solutions in the abstract" (2010: 35). By looking at a matrix we can map the space of possible solutions, or what kind of actions may become stable behavioral regularities in a given set of circumstances. But predicting which of these solutions will be implemented is a much more difficult task.

When game theorists map the space of possible outcomes, they look for Nash equilibria. Since this concept will be used repeatedly, it is important to understand its essential characteristics:

> *Nash equilibrium* (def.): a profile of strategies (or outcome) such that no player has an incentive to change her strategy (or action) unilaterally.

The last term—"unilaterally"—is important: in equilibrium each player has no reason or motive to change her strategy *provided that the other players do not change their strategies* either. If all the others do their part, no one has an incentive to deviate. The underlying intuition is that an equilibrium is a stable state: unless some parameters of the game change, each player is better off choosing her equilibrium strategy. As already mentioned, this does *not* mean that the players like that particular equilibrium. In fact it often happens (and we shall see some examples soon) that they would prefer some other outcome. But the concept of equilibrium implies that they cannot bring about another outcome by acting unilaterally.

On reflection, it is obvious that Nash equilibria must play some role in the analysis of institutions. Institutions usually facilitate the achievement of outcomes that individuals would be unable to bring about if they acted individually, in an uncoordinated fashion. Using Nash equilibria thus we can represent precisely those situations where institutions seem to be needed.

In this book I will focus almost exclusively on Nash equilibria "in pure strategies," or in other words, on strategies that do not involve randomized behavior. In the game of Figure 2.1, RR is the only Nash equilibrium in pure strategies. One way to find out is to pick a cell at random and ask whether one of the players would prefer to change her strategy. For example, if we pick LL, clearly Player 2 (column) would prefer to shift from L to R. Repeating this reasoning, we can see that LR cannot be an equilibrium either: Player 1 (Row) would prefer to switch from L to R. The only pair of strategies that is stable, because no one has an incentive to deviate unilaterally, is RR.

Not all games have a unique solution unfortunately. Many games have multiple equilibria, or several possible stable outcomes. This raises the so-called *problem of equilibrium selection*, and we will see that institutions can contribute to its solution. But before we come to that, let us clarify a few things: first, the concept of equilibrium presupposes only that the players react to incentives. It does not presuppose that equilibria are discovered through reasoning, or that the players solve games by thinking strategically about each other's actions. Strictly speaking, it does not even presuppose that the players can form representations (beliefs) of other people's preferences, expectations, and actions. This is particularly handy when we use games to model the interactions of nonhuman players, such as animals or organizations (e.g., firms). But when it makes sense to attribute beliefs to players, equilibrium states have interesting epistemic properties. In a Nash equilibrium, to begin with, the actions and the beliefs of every player are consistent: each player chooses her best action given her beliefs about the actions of the other players. Moreover, all beliefs are correct. For example, Player 1 chooses R because she believes that Player 2 chooses R; and Player 2 chooses R because she believes that Player 1 chooses R. The actions of each player are consistent with her own preferences and beliefs, and the beliefs of the various players are mutually consistent and true.

In the course of the book I will use frequently a small number of games with multiple equilibria, called *coordination games*. They are represented in Figure 2.2. Each matrix represents a class of games with similar payoff structures. Notice that the payoff numbers are not particularly significant by themselves. What really matters is the relationship between the numbers, or their structure. For example, if we multiply every number times three in the first matrix, we still obtain the same type of game (a driving game). As long as the payoff ranking is preserved, the fundamental strategic properties of the game do not change.

The labels are meant to recall the essential features of each kind of game, and are related to a stylized scenario or simple story.

> *Driving game*: Two drivers approach one another in a road with two lanes. They can either keep left or keep right. They are both better off if they both choose the same side, otherwise they will have to stop, wasting precious

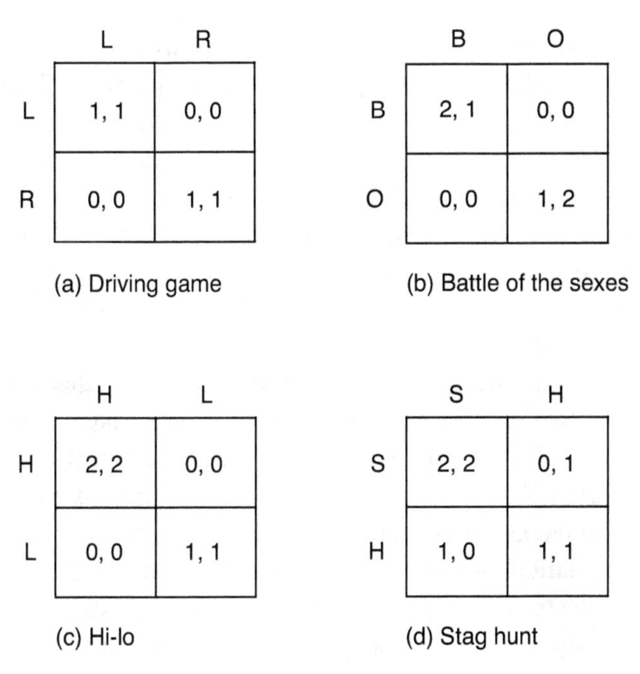

FIGURE 2.2. Four coordination games.

time, or crash. This matrix can be used to describe those situations where the players must join forces to solve a problem, there are many ways to solve the problem, and the players are indifferent about which way they do it. In particular, there are no conflicts of interest because both players benefit equally regardless of which equilibrium they choose.

Battle of the sexes: The story involves two fiancées who prefer to spend the evening together, but have different preferences concerning the available options. Row would prefer to attend a basketball game, while Column would prefer to go to the opera. This is a *bargaining* game, because the benefits of coordination are allocated in different ways, and some allocations are more favorable to one of the players. Battle of the sexes is useful to represent situations where there are conflicts of interest as well as benefits from coordination. In a military operation, either I lead and you follow, or you lead and I follow, but it would be disastrous if we all led or all followed. In a marriage either you cook and I wash the dishes, or I cook and you do the washing up; and perhaps we both prefer to cook. So even though we all agree that it is desirable to converge on an equilibrium, we are not indifferent about which one of the equilibria is implemented.

Hi-lo: The strategic setting is the same, except that one of the two options is clearly better for both players. Although hi-lo may appear a trivial game, it is in fact quite important for the study of institutions. Institutions facilitate convergence on an outcome that is superior to lack of coordination. But occasionally institutions can have negative effects, trapping people in an inferior stable state (LL). One may wonder how people can fail to see that (LL) is a silly outcome. One answer is that in some cases they simply do not know that there is a better equilibrium. In other cases the existence of a superior equilibrium is known, but the players fail to converge on it because they do not believe that the other players will do their part. This may happen, for example, when the inferior equilibrium strategy is safe and the superior one is risky, as in the "stag hunt" game (Figure 2.2(a)).

Stag hunt: In his *Discourse on the Origin of Inequality* (1755) Jean-Jacques Rousseau tells the story of a group of hunters trying to ambush a deer in the forest. The hunt requires coordination, discipline, and trust, for the players cannot see each other and the forest is full of distractions. Rousseau's outlook is pessimistic: "If it was a matter of hunting a deer, everyone well realized that he must remain faithful to his post; but if a hare happened to pass within reach of one of them, we cannot doubt that he would have gone off in pursuit of it without scruple."

	C	D
C	2, 2	0, 3
D	3, 0	1, 1

FIGURE 2.3. The prisoner's dilemma.

Chasing hare is tempting because it guarantees dinner whatever the other player may do. Hunting deer of course may bring a larger meal, but it is riskier. If the other hunter chases hare, the deer hunter will stay hungry. Formally the stag hunt game is identical to hi-lo, except that the out-of-equilibrium payoffs are larger for those who chase hare (H). Chasing stag is worthwhile only if the other player collaborates.

Tinkering with the payoffs is a good learning exercise. The stag hunt is basically hi-lo with asymmetric out-of-equilibrium payoffs. But what happens if we raise the out-of-equilibrium payoffs even more? As the temptation to deviate increases, the superior equilibrium becomes less and less attractive. Until at some point it ceases to be an equilibrium at all. This is represented in Figure 2.3: the stag hunt has become a "prisoner's dilemma" game.

The prisoner's dilemma is the most famous model in game theory. Its nickname comes from a story involving two robbers and a cunning sheriff. In one of the earliest textbooks of game theory, the story was told like this:

> Two suspects are taken into custody and separated. The district attorney is certain that they are guilty of a specific crime, but he does not have enough evidence to convict them at a trial. He points out to each prisoner that each has two alternatives: to confess to the crime the police are sure they have done, or not to confess. If they both do not confess, then the district attorney states he will book them on some very minor trumped-up charge such

> as petty larceny and illegal possession of a weapon, and they will both receive minor punishment; if they both confess they will be prosecuted, but he will recommend less than the most severe sentence; but if one confesses and the other does not, then the confessor will receive lenient treatment for turning state's evidence, whereas the latter will get "the book" slapped at him. (Luce and Raiffa 1957: 95)

In the standard matrix, D stands for "defect" and corresponds to the prisoners' decision to confess. C stands for "cooperate" and corresponds to the decision to keep their mouths shut.

The prisoner's dilemma is a very peculiar kind of game, and must not be confused with the games we have examined so far. The driving game, hi-lo, and the stag hunt game have multiple equilibria. They are *coordination* problems. The prisoner's dilemma is different because the outcome in the top-left corner is *not* an equilibrium: each player is better off if she plays D unilaterally. This is puzzling in several respects. Recall that in the stag hunt each player has the problem of guessing the move of the other player; in the prisoner's dilemma that problem does not even exist: the temptation to defect in a sense has become so strong that she does not have to think about the action of the other player. Whatever she does, it is better to play D. This means that in the prisoner's dilemma there is only one equilibrium (DD), which is inefficient. To distinguish, we shall say that this kind of game represents a problem (or dilemma) of *cooperation*, as opposed to *coordination*. This stretches the ordinary meaning of "cooperation" a bit, but it is useful to have a distinct term for each case.

In spite of this analysis, many people think that cooperation is the right choice in the prisoner's dilemma game. Why? There is probably more than one reason, including a certain difficulty to think strategically on the part of many people. But first and foremost, people have this intuition because in real life they are accustomed to follow rules that prescribe cooperation in situations that are similar to the prisoner's dilemma. There is a crucial difference though: the rules of real life typically prescribe *conditional* cooperation: cooperate if the other player cooperates, otherwise defect. This works well when we play a game repeatedly and we can see how the other players behave in similar situations. It does not matter whether they have played with us or with someone else in the past; the important point is that we can check if they have been cooperative. Based on this information, we can then decide to cooperate or defect in the next encounter, and the threat of future defections is often sufficient to induce the other player to cooperate as well.

It is possible to demonstrate that in certain conditions conditional cooperation is an equilibrium of the prisoner's dilemma game when it is played an indefinite number of times. But it can also be shown that many different rules of conditional cooperation can be sustained by the threat of defection. We may follow a rule that says that I cooperate unilaterally on Monday, Tuesday, and Wednesday, and you cooperate unilaterally for the rest of the week, for example. Even though this may seem unfair, it would still be preferable to choosing mutual defection every single day of the week. But there is also an equilibrium where *you* cooperate on Monday, Tuesday, Wednesday, and I do it for the rest of the week. So repeated encounters facilitate cooperation, but at the same time they raise the problem of coordination, by multiplying the number of possible equilibria. (Remember battle of the sexes? I wash and you cook, or I cook and you wash, or I cook two days a week, or three, or four. . . .)

Earlier I have quoted Schelling's view that game theory is better at mapping the territory than at providing specific explanations of specific behaviors. Part of the problem is that some games have a large number of equilibria, and we do not have a good general theory of equilibrium selection. By "theory" here I mean something like a "mechanistic explanation," a detailed account of the causal processes that induce the players to converge on an equilibrium solution. But even without such a theory the identification of equilibria is an important achievement, because equilibrium models help the formulation of *functional explanations* of behavior. The general form of a functional explanation is "X (exists or occurs) because it causes Y." A familiar example is the explanations of biological traits that appeal to the benefits the trait confers to an organism ("the heart exists because it causes blood to circulate in the body," for example). But functional explanations are also common in the social sciences. Standard textbook expositions of monetary theory, for example, begin with a functional analysis of money as medium of exchange, store of value, and unit of accounting.

Philip Pettit (1996) has noticed that functional explanations may be used to explain the emergence of a trait or behavior, or its resilience in given circumstances, or both. Now, equilibrium models are particularly suitable to formulate functional explanations of resilience. Suppose for instance that some people have found a set of rules that, if followed, help achieve a result that they could not achieve if they acted independently. Suppose moreover that everyone has an incentive to comply with the rules if everyone else does the same. Although an equilibrium model is unable to tell us exactly how they have discovered these rules, it is able to explain why people continue to follow them even in spite of small deviations, and what must happen (how the incentives must change, how the beliefs must be manipulated) to make them stop following the rules. Thus the equilibrium model will support a functional explanation of the form: "the rules exist because they help people solve a coordination problem."

For example, suppose that some drivers have discovered that following a rule like "keep left" reduces the number of accidents and makes the traffic flow smoothly. Or suppose that a group of traders have found that exchanging paper certificates makes business transactions easier than carrying around golden bars or coins. In such cases, it will be natural to say that the rules of traffic and paper money fulfill important social and economic functions: they help people attain better outcomes; the outcomes are stable as long as the incentives do not change significantly. The rules persist because they are equilibria of strategic games.

Equilibrium models are particularly valuable when a behavioral regularity is supported by a variety of different causal mechanisms. In the case of traffic rules, for example, drivers tend to follow them because it is in their interest to do so, because they expect the other drivers to do the same, because they believe that they ought to, because the police fine transgressors, and just out of habit, because they are used to it. In such circumstances, any theory that focuses on the proximal causes would be extremely complicated. Scientists thus prefer to take a step back and model the situation as a system that for a variety of reasons tends to remain in a given state—the equilibrium state.

Throughout the book I will follow Pettit and other theorists who hold that equilibrium models provide legitimate explanations of the persistence of behavioral patterns. This does not mean of course that equilibrium models are well suited to provide other types of explanation, or that persistence is the only kind of phenomenon worthy of investigation in the social sciences. On the contrary, there are many other interesting aspects of social reality that call for different styles of explanation. But simple game-theoretic models can illuminate many features of institutions that otherwise would appear rather incomprehensible. In the chapters that follow we will begin to see how.

REFERENCES AND FURTHER READINGS

There are many excellent introductory textbooks of game theory; my favorite ones are Dixit, Skeath, and Reiley (2009) and Osborne (2004). The main equilibrium concept in game theory is named after John Nash, the troubled mathematician portrayed in the Hollywood blockbuster *A Beautiful Mind.* Bicchieri (2001) discusses in depth the problem of belief formation in coordination games. The idea that society is a gigantic coordination game and that social institutions help people find a solution is already in Hume (1748), but has been reformulated in game-theoretic fashion by philosophers and social scientists like Lewis (1969), Ullmann-Margalit (1977), Schotter (1981), Sugden (1986), Skyrms (1996, 2004), Binmore (1998, 2005). Knight (1992) discusses the coordinating function of institutions, as well as the competitive aspect of games such as battle of the sexes. The puzzle of cooperation in prisoner's dilemma experiments has generated an enormous literature, but explanations based on social norms are quite common—see for instance Ullmann-Margalit (1977), Sober and Wilson (1998), Binmore (2005), Bicchieri (2006), Gintis (2009). Functional explanations have suffered from a bad press in social science as a consequence of Jon Elster's (1983) influential critique. Pettit's (1996) distinction between explanations of emergence and explanations of resilience is meant to resist this critique. On equilibrium explanations, finally, see Sober (1983).

CHAPTER 3

MONEY

Money is a fundamental economic institution. It is also a familiar thing, part of the social world as it appears to us in everyday life. And yet, it is mysterious and fascinating. Philosophers and social scientists have theorized about money for centuries, and over time money has become a key test case in social ontology. A theory that cannot explain money satisfactorily is not taken seriously in contemporary philosophical debates.

In this chapter I will briefly illustrate how social scientists' conception of money fits the account of institutions that I have sketched in chapter 1. In particular, I will explain in what sense money can be considered an equilibrium solution to a problem of coordination. Following the work of J. P. Smit, Philip Buekens, and Stan du Plessis, I will emphasize the role played by incentives. If institutions are rules, they must be rules that people are motivated to follow. Although this chapter is mostly descriptive and does not include any original argument, it may be interesting to those philosophers who think that social scientists have a poor understanding of the ontology of money. John Searle, for example, has criticized economists harshly:

> When I studied economics as an undergraduate at Oxford, none of my teachers worried about the ontological presuppositions of the investigation. . . . I think it is sometimes

> possible to do good research without worrying about the ontological issues, but the whole investigation gets a greater depth if one is acutely conscious of the ontology of the phenomena being investigated. It is, for example, a mistake to treat money and other such instruments as if they were natural phenomena like the phenomena studied in physics, chemistry, and biology. The recent economic crisis makes it clear that they are products of massive fantasy. As long as everyone shares the fantasy and has confidence in it, the system will work just fine. But when some of the fantasies cease to be believable, as happened with the subprime mortgage instruments, then the whole system begins to unravel. (Searle 2010: 201)

This rebuke deserves to be taken seriously. Do economists really ignore what money is? Are economists unaware of the importance of collective beliefs ("fantasies") and the role they play in finance? If true, it would be a serious flaw of their theories.

So, suppose we wanted to know more about economists' conception of money: where should we look? Unfortunately you do not find an "ontological" analysis of money simply by opening a macroeconomics textbook. Textbook economics focuses on the relations that link money with other economic variables like prices, investments, and growth. This is typical of most science: you do not find an in-depth analysis of the nature of gravitation in a Physics 101 course, for example. Introductory textbooks illustrate models that relate gravitation to variables like speed, acceleration, and mass, without explaining what gravitation really is. But this does not mean that physicists have nothing to say about the ontology of gravitation. It just means that you have to look elsewhere, in an advanced text of quantum field theory for example.

The same is true of economics. Although Searle did not see much of it in the introductory courses that he attended in Oxford in the 1950s, economists have a perfectly sound conception of the ontology of money. Nobuhiro Kiyotaki and Randall Wright, two leading contemporary theorists, for example write at the outset of one of their classic articles that "a critical factor in determining if an object can serve as a medium of exchange is whether or not agents believe that it will" (1989: 928). And in their Clarendon Lectures, Kiyotaki and John Moore point out that for money to work as a medium of exchange

> there has to be a set of mutually-sustaining beliefs, stretching off to infinity. I was willing to hold money yesterday because I believed the dentist would accept it today. She is willing to hold money today because she believes someone else will accept it tomorrow. And so on. If there were a known end-point to history, the entire structure of beliefs would collapse back from the end. (2001: 3)

So it seems that economists are aware that money is the effect of a "massive fantasy," to use Searle's colorful language.

Economists have studied the ontology of money since the nineteenth century at least. In fact it turns out that they do not just have one theory of money—they have many. These theories deal with issues such as the creation of money, its resilience, and its reliance on systems of mutual expectations. Economists' analyses usually begin with a functionalist definition of money: *money is what money does.* The functionalist definition suggests that economists are aware of the conventional character of money: money is whatever fulfills the various functions of money, irrespective of which particular medium is used in a specific context. Shells, fur, pieces of metal, paper, electronic bytes can all be money if they fulfill the functions of money.

What are these functions, then? Most textbooks mention three. Money works as a medium of exchange, unit of accounting, and store of value. These functions are related in interesting ways: if a currency is used as a medium of exchange, for example, then it can be used as a unit of accounting too (if one has to pay twice as much to obtain A than to obtain B, then the market value of A is twice the value of B). Moreover, a currency can be used as a medium of exchange only if its value does not decay too quickly during the time elapsing between the sale of commodity A and the purchase of commodity B. The medium of exchange function thus presupposes a certain capacity to work as a store of value. Notice the importance of *time*: it suggests that expectations about the future (that is, beliefs) play an important role in economic theories of money.

Scientific theories of money are usually classified in two grand categories, called commodity and claim theories. Commodity theories of money go back to the work of Karl Menger, one of the pioneers of the marginalist approach that still constitutes the bulk of contemporary economics. His essay "On the Origin of Money" (1892) was simultaneously a theoretical and a historical reconstruction of the emergence of the institution of money. Its historical inaccuracy is generally recognized today, and so it is only taken as a rational reconstruction of how money could or would have evolved in idealized circumstances. Menger's story however is still considered useful to understand what money is and how it functions, if not where it comes from.

Menger's story is articulated in four steps:

1 Consider a society with a primitive form of division of labor. Division of labor improves efficiency, but at the same time creates the problem of the *double coincidence of wants*. Imagine that three individuals (Ann, Bob, and Carol) specialize as follows: Ann produces meat, Bob produces vegetables, and Carol produces fruit. Now suppose that Ann wants to consume some vegetables, but Bob is not interested in meat; rather, he would like to have some fruit. In principle Bob could obtain some meat temporarily from Ann, and exchange it later with Carol to get some fruit (assuming that the latter is interested in meat). Meat, in this simple example, works as a medium of exchange.

2 As this process is repeated again and again, a single medium is likely to emerge to facilitate exchange. Some commodity may be used more frequently than others as a medium because of its special properties: unlike meat, a good medium of exchange must be durable, portable, divisible, and universally desired. A classic example is gold, a metal that displays most of these characteristics. Historians point out that in reality the media that emerge in early market societies are rarely unique, and that some societies use for long periods of time several commodities simultaneously. But as I said, what is interesting is the basic mechanism, not historical accuracy.
3 The third step consists in replacing the physical commodity with paper certificates. Gold is stored in safe places (banks) that issue certificates stating that the bearer owns a certain amount of gold. Instead of exchanging gold (which can remain in the banks), people prefer to exchange certificates. Notice that this already requires a system of *trust*, that is, people must believe that banks are reliable, that they do not cheat, and that everyone else believes the same. If this condition does not hold, it is pointless to carry certificates, because no one else would accept them as payment. That certificates are used as currency must be commonly believed in the relevant community.

Menger's account is an "invisible hand" story. Money is not invented by a central authority, but emerges spontaneously from repeated interactions between individuals, to facilitate market transactions. Although invisible hand stories are the bread and butter of social science (as well as biology) several scholars have found faults with Menger's story, for two reasons. One, that I have already mentioned, is historiographical: money does not seem to have emerged this way. (With a bit of patience we shall see why.) The second reason is that an invisible hand story faces difficulties when it tries to account for the fourth step in the development of money.

4 Gold is done away with and replaced by *fiat money.* Central banks issue certificates that do not grant the ownership of an amount of gold, but only refer to an abstract currency (sterling, dollar, franc, euro, lira). And yet, surprisingly, people use the certificates—paper money—as a medium of exchange. They accept them for payment, exchange them for goods, even though they do not grant the ownership of any underlying commodity.

This is a problem for Menger because fiat money has no value in itself; so why should people want to hold it? The answer is: because they believe they can use it to purchase other goods in the future.

But then they must believe that other people will accept fiat money in the future, or that these people will believe that other people will accept it in the future, and so forth. A system of consistent and mutually sustaining beliefs is required.

But where do these beliefs come from? And how are they reproduced? Are they not too fragile? It seems that if someone begins to doubt the future viability of the currency, the whole system of expectations will crumble. The main problem of an invisible hand theory is to give a plausible explanation of why such systemic crises are relatively rare, rather than the norm as one would expect if the currency is not backed up by gold.

We know already that self-sustaining systems of beliefs may be backed up by past regularities of behavior: if everyone has used this kind of paper as a medium of exchange up until now, I might use past behavior as a signal that others will continue to use it in the future, that they make the same inference, and so forth. This is basically Lewis's story, as we have seen. But surely beliefs can change. For example, in periods of uncertainty peo-

	Currency	Gold
Currency	2, 2	0, 1
Gold	1, 0	1, 1

FIGURE 3.1. Fiat money as a stag hunt game.

ple may prefer not to hold paper certificates (except for immediate needs) and ask for other commodities as payment (gold, cigarettes, fuel, or a more stable foreign currency). If I know this, I will also prefer not to be paid in paper certificates, which will lead to further devaluation and so on.

This is known as the "free riding problem of fiat money." Strictly speaking it is not a prisoner's dilemma game. If everyone else prefers to use fiat money rather than barter, then I also prefer to use fiat money (there is no point in asking for commodities or gold). But if I am not too sure about what others will do in the future, then it may be wise to hold commodities rather than fiat money. The strategic structure is similar to a stag hunt game (Figure 3.1).

We know that stag hunt games have multiple solutions. The key to avoid the inferior equilibrium state (Gold, Gold) is to convince everyone that they *will* need currency in the future, and this is something that only a powerful enforcing authority can do. At this point we ought to abandon the pure commodity theory of money and move to its main rival, the so-called *claim theory* (or chartalist, or state theory). The core idea is that money is created and sustained by state authority. The state intervenes by means of two mechanisms:

(i) first, the state pays civil servants with vouchers,

and then

(ii) it forces all citizens to pay taxes using these vouchers.

The second step is crucial: as a consequence of the state's request, everybody must hold some vouchers (fiat money), even though the vouchers are not desirable in themselves. What makes them valuable is the stamp of the state, which collects them back through taxation. But a key precondition for this mechanism to work is the state's credibility as a source of power.

If the state is shaky, it may not be in my interest to hold any fiat money: in six months' time there may be another authority in place, asking me to pay taxes by means of another currency. Notice that the state must also be a credible supervisor of the quantity and quality of money. Politicians may be tempted to print an increasing quantity of money to buy civil servants. But if the latter do not produce enough commodities and services, the flood of vouchers will generate inflation. Inflation in turn discourages people from holding currency, and in the long run may cause the entire monetary system to collapse. Only a strong, stable, and serious state can back up the system of self-sustaining beliefs with enough credibility to make a fiat currency viable.

According to historians and anthropologists this is a more accurate portrait of the emergence of money. But the claim theory also has a theoretical advantage: it explains how the beliefs that make money robust as an institution may become resilient. A currency is robust as long and insofar as the authority that issues the currency (the state) is a credible enforcer. The state acts in part as a coordination device, by stating rules that make one profile of strategies salient out of many possible coordination equilibria ("use *this* particular currency as a medium of exchange"). But it also makes use of its power to provide incentives and to punish deviations from the fiat money equilibrium. People incur harsh penalties if they do not pay taxes or if they issue counterfeit money.

So fiat money is sustained by two mechanisms: (1) individuals know that fiat money reduces their transaction costs, and (2) they know that they will be punished if they do not use it in some transactions (e.g., tax payments). The Venetian traveler Marco Polo highlighted these mechanisms in his diaries, when he described the mint of the Great Kahn in China—an extraordinary institution for a thirteenth-century European merchant:

> In this city of Kanbala is the mint of the Great Kahn, who may truly be said to possess the secret of the alchemists, as he has the art of producing money by the following process . . . the coinage of this paper money is authenticated with as much form and ceremony as if it were actually pure gold or silver; for to each note a number of officers, specially appointed, not only subscribe their names, but affix their seals also . . . in this way it receives full authenticity as current money, and the act of counterfeiting it is punished as a capital offence. When thus coined in large quantities this paper currency is circulated in every part of the Great Kahn's dominions; nor dares any person, at the peril of his life, refuse to accept it in payment. All his Majesty's armies are paid with this currency, which is to them the same value as if it were gold or silver. Upon these grounds, it may be certainly affirmed that the Great Kahn has a more extensive command of treasure than any other sovereign in the universe. (Polo 1298: 95)

The claim theory is often considered alternative to Menger's commodity theory, because the two theories use different styles of explanation: one is an invisible hand theory, the other appeals to central authority. The two theories give radically different accounts of the emergence of money, but the underlying ontology is the same. In fact for our purposes their similarities are more significant than their differences: both emphasize the role of mutually consistent beliefs. And both appeal to an equilibrium type of explanation: every citizen has an incentive to use the currency, provided that the others do the same.

So it is false that contemporary economists ignore the subtle ontological issues raised by the existence of money. They have a perfectly sound account of the nature of money, which is consistent with the contemporary scientific analysis of social institutions that I have sketched in the previous chapters. Moreover, such accounts are also largely consistent with what philosophers say about money: the dependence of money on collective beliefs is hardly a novel insight, and one should not rebuke social scientists for ignoring a view that they have held for many decades, if not centuries.

REFERENCES AND FURTHER READINGS

Smit, Buekens, and du Plessis (2011) illustrate economists' standard conception of money as a system of beliefs sustained by individual incentives, and contrast it with the conception of philosophers like Searle (1995, 2010). Tieffenbach (2010) compares Searle's account with Menger's commodity theory. For a contemporary formal analysis of money in the tradition of the commodity theory, see for example Kiyotaki and Wright (1989, 1991). The claim theory is more like a family than a single theory. Early formulations can be found in the writings of Max Weber (1922) and Georg Friedrich Knapp (1905), but the theory still has many disciples today. See for example Goodhart (1989) and Wray (1990). Pryor (1977) offers some empirical evidence about taxation, state authority, and the origins of money. Austrian economists, following Hayek (1976), reject the claim that the power of the state to levy taxes is crucial to stabilize fiat money, and advocate competition between different currencies issued by private banks. This view, however, is minoritarian in contemporary economics. The English translation of Marco Polo's description of the mint of the Gran Kahn is taken from Smit, Buekens, and du Plessis (2011: 20).

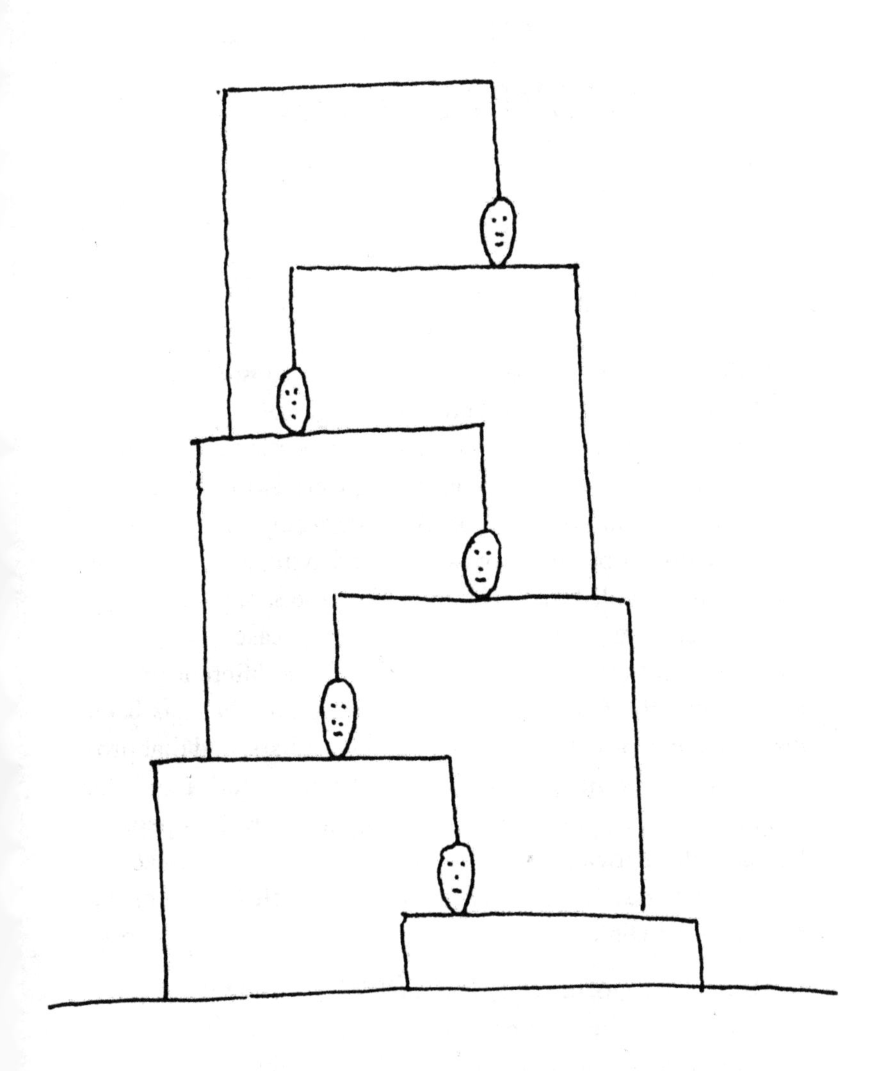

CHAPTER 4

CORRELATION

We have seen in chapter 1 that conventions can be modeled as solutions to coordination games with multiple equilibria. Lewis's analysis focused on games with symmetric payoffs, where the players do not strongly prefer to converge on one rather than another solution. A classic example is the driving game: we do not particularly care about driving on the right or on the left, provided that everybody does the same. The theory however can easily be generalized to other cases, where the payoffs are asymmetric and the players have different preferences about the outcomes. Here is an example that has been discussed in depth by scientists and philosophers, and that provides an interesting story (or "fable" if you prefer) about the origins of private property. The protagonists are two peoples living in the African savannah. As a homage to the great anthropologist Edward Evans-Pritchard, we shall call them the Nuer and the Dinka:

> Many years ago, the Nuer and the Dinka settled in the valley of Sobat. The Nuer came from the north and the Dinka from the south, looking for green pastures for their cattle. Each tribe occupied as much grazing land as possible, until they arrived at the banks of the river Sobat. Moving their cows across the river would have been difficult, so each tribe grazed on one side of the river only (Figure 4.1a).

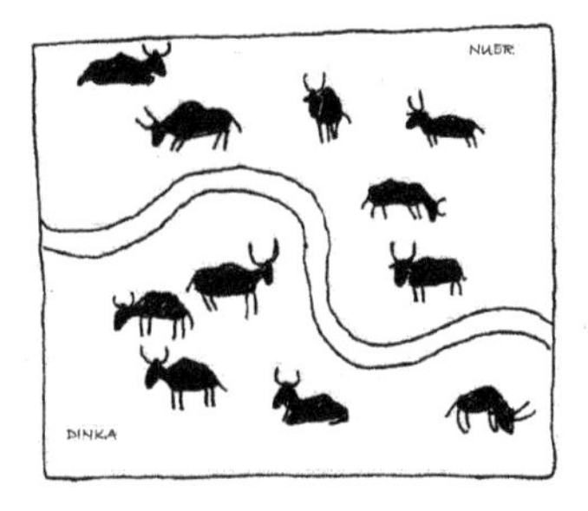

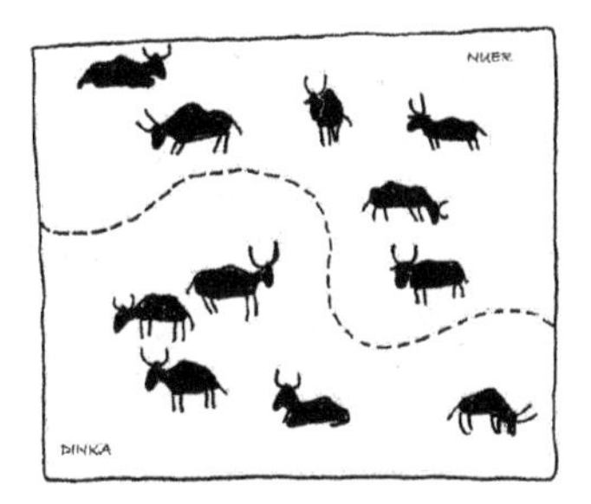

FIGURE 4.1. Equilibrium in the valley of Sobat.

> Over the years the river progressively lost its water, due to changes in the region's climate, until at one point it became completely dry. Only a sandy line separated the areas occupied by the Nuer and the Dinka. The members of the two tribes could now easily trespass the old river's bed, and graze their cattle wherever they wished. But each piece of land now could be contested, and conflicts could easily escalate in outright war (Figure 4.1b).

The grazing game of the Sobat Valley can be represented in strategic form using a matrix known as "hawk-dove" in biology, and "chicken" in economics. Every time a Nuer and a Dinka man spot a new piece of land, they have to make a decision: in Figure 4.2 the strategy G stands for "graze," NG for "not graze." If they decide to graze the same area, the members of the two tribes will end up fighting, which is the worst outcome for all (0, 0). If they both abstain, they will not clash but will miss the opportunity to feed their cattle (1, 1). The best solution is to converge on one of the two equilibria in the top-right and bottom-left corners, where the member of one tribe grazes and the other lets him graze. But who is going to give way?

The grazing game is a problem of coordination with asymmetric equilibria, depending on who is going to give way. But since the players are identical, why should one of them accept a

	G	NG
G	0, 0	2, 1
NG	1, 2	1, 1

FIGURE 4.2. The grazing game (hawk-dove).

lower payoff? Notice that the only symmetric solutions here are not only inefficient—they are not even equilibria of the game. As a consequence we should expect some player to deviate unilaterally sooner or later.

Our fictional story was designed so as to make a solution stand out as "obvious." The story continues like this:

> Although it would have been easy to trespass into the other tribe's territory, the Nuer and the Dinka preferred to avoid conflicts. The Nuer kept grazing on the north side, the Dinka on the south of the old river's bed (Figure 4.1b). The strip of sand could not physically stop raiders, but each tribe was happy to treat it as a border dividing their territories.

Borders and territories are institutional entities. The Nuer and the Dinka have developed a rudimentary institution of private property. But how can this institution be modeled in game-theoretic form? Notice that borders and territories are not even represented in the grazing game (Figure 4.2): they are extra-theoretic features that help find a solution to the coordination problem. As opposed to the actions in Figure 4.2, the solution devised by the Dinka and the Nuer involves a new set of *conditional* strategies. Each tribesman, in particular, conditions his move (G or NG) on the position of the patch of land relative to the old river's bed. If it is north the Nuer graze it, if it is south only the Dinka do it. This is similar to the solution of other problems analyzed by Lewis, such as the driving game that I have mentioned a couple of chapters ago. In that case, the drivers condition their choices on the history of play: if everybody has been driving on the right up until now, everybody will continue to drive on the right; if everybody has been driving on the left, it is in everybody's interest to keep doing the same. The only difference is that the conditional strategy in the road game does not lead to a substantially different outcome than any of the two unconditional strategies ("keep right," "keep left"). In the grazing game it does, for none of the unconditional strategies can deliver symmetric payoffs. This capacity—the capacity to create new outcomes—is an important feature of many institutions, as we shall see shortly.

What kind of equilibrium can a convention be? Although there are two Nash equilibria, none is a convention in the grazing game. As a consequence, conventions cannot be just Nash equilibria of coordination games. Peter Vanderschraaf (1995) has shown that Lewis's conventions are *correlated equilibria*, a solution concept first studied by Robert Aumann in the 1970s. Correlated equilibria play an important role in the unified theory that I propose in this chapter, so it is important to have an intuitive understanding of their characteristics. Since the formal models are somewhat complicated, I will provide an informal account and leave interested readers to pursue the details in the technical literature.

To grasp the idea of correlated equilibrium, it is useful to start from a hypothetical preconventional scenario. Suppose the Dinka and the Nuer are about to play the grazing game, and that (by hypothesis) no solution is salient. In such circumstances their only option is to choose randomly. The Nuer decide to toss a coin: if head comes up they will choose G, if tails they will choose NG. The Dinka decide to do the same, and toss their own coin. Combining the probabilities that they obtain different results, they have a 50 percent chance to converge on one of the efficient solutions. Their expected payoffs, unfortunately, are not larger than those that result from never grazing the contested land (1, 1).

The coins in this example have been tossed *separately* and *privately*. Would it make any difference if, instead, the coin toss was a *single* and *public* event? Let us introduce a new character: following Herbert Gintis (2009) I shall call it the "choreographer." The choreographer tosses the coin and announces publicly: "the Nuer graze" if head, or "the Dinka graze" if tails comes up. The two players see the coin-tossing ceremony and know that the other can see it too. Moreover, they know that the other knows (and so forth) that they both see the same ceremony: the outcome of the toss is common knowledge.

In such circumstances, it would seem reasonable to follow the advice of the choreographer. Each player, in other words, could condition her behavior on the result of the toss, following the obvious strategy: "choose G if the choreographer says so, otherwise choose NG." If both players are confident that the other one is going to follow this strategy, then they are better off tossing the coin publicly. They would always coordinate on an efficient outcome, exploiting the results of the ceremony.

Solutions of this kind are correlated equilibria. A correlated equilibrium of a game G is a Nash equilibrium of a larger game G* obtained augmenting G with the addition of new strategies. The new strategies prescribe actions conditional on the occurrence of an external event that is not part of the original game. They take the form of "if X then do Y" statements, where X is a property of the correlation device. As Vanderschraaf has shown, Lewis's conventions are correlated equilibria that exploit earlier moves in the coordination game. The coin, in other words, is replaced by the history of play.

What would an augmented version of the grazing game look like? The introduction of an external device breaks the stalemate, creating a better solution than those that were previously available. The external device that the Nuer and the Dinka use to condition their strategies is the location of the contested piece of land with respect to the river Sobat. If the land is north of the river, the Nuer graze it; if it is south, the Dinka graze it. This is

	G	NG	G if S, NG if N
G	0, 0	2, 1	1, ½
NG	1, 2	1, 1	1, 3⁄2
G if N, NG if S	½, 1	3⁄2, 1	3⁄2, 3⁄2

FIGURE 4.3. Augmented grazing game.

analogous to tossing a coin or using past behavior as a guide for future action in the driving game. We construct an augmented grazing game adding two conditional strategies to the game in Figure 4.2 (with "N" for "north" and "S" for "south"):

(i) G if N, NG if S
(ii) G if S, NG if N

Again let us assume for simplicity that in each instance there is 50 percent chance that the land lies north (or south) of the river. Notice that in the augmented game (Figure 4.3) the outcome of the conditional strategies is the only symmetric Nash equilibrium in pure strategies (3/2, 3/2). The other two equilibria (G, NG) and (NG, G) are still there, and each player would in principle prefer to converge on the one that is most favorable to her. But conditioning on an external event provides a third stable solution that should be acceptable to both.

Of course one may ask: Why *this* kind of correlation? Why do they pay attention to the river, rather than to some other element of the environment? Surely one can construct a thousand augmented games, corresponding to thousands of correlated equilibria that exploit different external events. "Graze if sunny, do not graze if cloudy"; "graze if Monday, do not graze if Tuesday": these are all potential correlated strategies. Even "always graze if you are Nuer, do not graze if you are Dinka" (G, NG) is potentially a viable institution. So why "G if S, NG if N"? What is so special about it? Schelling and Lewis argued that a strategy profile may be selected in virtue of its salience. That the Nuer graze north and the Dinka graze south of the old river is salient in the Sobat Valley because it had been the traditional arrangement for many years, before the river became dry. History creates a focal point for coordination. If no event or correlation stands out "naturally," then a group of ingenious players can try to make one of them *more* salient—by means of teaching, acculturation, and training. But there is no game-theoretic account of which correlations we naturally hook onto. Social ontology merges with social history, psychology, and biology at this point, and theoretical speculation must give way to empirically informed models. Clearly the theory of correlated equilibria does not solve the "mystery" of salience. It does, however, solve an important puzzle of social ontology. Recall the question we started from: are institutions *rules*, or *equilibria* of a game? We can now see that the answer is "both": an institution may be considered as an equilibrium or as a rule of the game, depending on the perspective that one takes. The correlated equilibrium in the game of the Sobat Valley is the pair of strategies:

(i) G if N, NG if S
(ii) G if S, NG if N

From the point of view of an external observer, the grazing convention of the Sobat Valley takes the form of a *regularity* that corresponds to the correlated equilibrium in G (or, which is equivalent, to the corresponding Nash equilibrium in G*). But each strategy in this profile also takes the form of a *rule* that dictates each player what to do in the given circumstances. The Nuer therefore will perceive the institution as a prescription to graze their cattle if the land is north, not to graze it if it is south. And, mutatis mutandis, the same is true for the Dinka (G if S, NG if N). Since the two strategies are formulated as rules, clearly the equilibrium is a set of rules—one for each player—that "establish a stable structure to human interaction" (North 1990: 6).

It is worth emphasizing again that it would be impossible to conciliate these two views if we focused on the Nash equilibria of the original matrix (Figure 4.2): the conditional strategies (rules) are not even part of this game. And they cannot be: there is no north/south correlating device in the original game. So it is correct to see the correlated strategies as *external* rules that help attain coordination in the original game, in the spirit of Douglass North. But the convention is not a Nash equilibrium of the original game—it is a correlated equilibrium of the old game, and a Nash equilibrium of the *augmented* game. The contrast between rules and equilibria approaches is perhaps due to a failure to appreciate this distinction between equilibrium concepts. But with the introduction of correlated equilibria both approaches are vindicated: we have achieved a unified view of social ontology. We shall call it *the rules-in-equilibrium* theory of social institutions.

The rules-in-equilibrium account unifies the two dominant approaches in the literature—the rule-based approach and the equilibrium-based approach to the analysis of institutions. We have seen in chapter 1 that the rules approach by itself is deficient, because it cannot distinguish between effective and ineffective rules, and leads to regress when we ask how people interpret the rules. But how about the equilibria approach? How does it differ from a rules-in-equilibrium theory of institutions?

There are various reasons to believe that a theory that simply equates institutions with equilibria cannot be right. Although institutions may be *particular* types of equilibria, it is clear that not all equilibria are institutions. Take mutual defection in a prisoner's dilemma game for example. The pair of strategies DD is an equilibrium, but intuitively it is not an institution. (In the game-theoretic literature, in fact, mutual defection is taken to

	C	D
C	2, 2	0, 3
D	3, 0	1, 1

FIGURE 4.4. A prisoner's dilemma game.

represent the prototypical failure of sociality.) Why? The reason is not merely that mutual defection is worse than mutual cooperation. The reason is that in the prisoner's dilemma each agent can implement the defection strategy independently. There is no need to correlate one's actions with those of the other player. In fact there is no reason to even think about the action of the other player: whatever she does, it is optimal to defect.

Let us stipulate that institutions are equilibria of *coordination* games, then. Although this condition takes us in the right direction, it is still too permissive. The problem is that correlated equilibria are too common: several nonhuman animals solve coordination games exploiting correlation devices, and animals do not have institutions.

A classic example is *Pararge aegeria*, a butterfly living in the woodlands of Asia and Europe. Male butterflies patrol the patches of sunlight that appear on the woodland's floor, where they mate with females after a brief courtship. When a male enters a sunspot that is already occupied by another male, it is attacked by the incumbent. After a brief fight, the defeated butterfly leaves the spot. Remarkably, the intruder is nearly always defeated, and the incumbent nearly always retains its territory. Similar patterns have been observed in swallowtails, baboons and lions. The standard interpretation is that males engage in repeated hawk-dove games, where they both have an interest to avoid the out-of-equilibrium outcome (hawk, hawk). As a solution, they have evolved pairs of strategies that minimize damage by granting the territory and the mating opportunity to the incumbent after a purely symbolic contest. The biologist John Maynard Smith, who first used game theory to explain such behavior, has called it "bourgeois equilibrium."

We are reluctant to speak of institutions in the butterfly case—except in a metaphorical sense. Many nonhuman animals are social creatures, but they live in societies without institutions. The expression "animal convention" seems to be more appropriate to describe these equilibria, where "convention" refers to the fact that evolution could have selected a different solution to the coordination problem. But whatever equilibrium has been selected, the important point is that the strategies are biologically implemented in each species. They are—taking the term "genetic" broadly—written in their genes. A group of butterflies cannot coordinate on anything but who occupied the sunspot first. They can play only this particular strategy. They cannot invent a new equilibrium. Humans, in contrast, can: they hook onto different correlations, invent constantly new strategies, and dramatically enlarge the number of possible equilibria.

Organisms can have many degrees of complexity, and of course some animals have more sophisticated cognitive capacities than insects. Some of them can learn by imitation, and some develop simple cultures. So we should take butterflies and humans only as ideal types, keeping in mind that continuity is the norm in nature. But what does distinguish human institutions from the correlated equilibria of *Pararge aegeria*? Butterflies react to a narrow set of signals only (who's in the spotlight first). A simple mechanism that links one type of stimulus with one type of behavior guarantees coordination. More complex creatures in contrast are able to decouple stimulus and behavior. They decouple by adding an intermediate state—a representation—that they can use to condition their behavior. Moreover, they can condition their strategies on *different* representations—they can use many signals and many correlation devices. In the case of humans, we say that they can follow different rules.

Strictly speaking, rules are not essential in standard equilibrium-based accounts of institutions. Equilibrium theories only require actions, strategies. Actions of course can be *expressed* or *described* by rules, that is, by means of statements such as

(i) G if N, NG if S
(ii) NG if N, G if S

I said earlier that a pair of actions like these (an equilibrium) may appear as a behavioral regularity to an observer, and as a rule to the participants. But this means that the concept of rule is ambiguous. We use rules sometimes to describe, sometimes to prescribe, and occasionally to do both things at once. These functions however are conceptually independent, so it is useful to introduce a distinction between two types of rules. Let us call them agent-rules (or *a-rules* for short) and observer-rules (*o-rules*) respectively: an observer formulates an o-rule mainly to represent or summarize others' behavior; an agent formulates an a-rule to represent and to guide her own behavior.

In standard equilibria accounts this guiding role is typically downplayed. Equilibrium theories are observer theories, and so the actions of the players are described from an external point of view only. This is not a flaw of game theory, but merely a consequence of game theorists' ambition to formulate an entirely abstract and general theory. Equilibrium accounts must be applicable to creatures with rather minimal cognitive capacities, such as butterflies. But game theorists are aware that the capacity to represent and to follow rules may help convergence on an equilibrium. An adequate theory of institutions thus must be more specific than an equilibrium theory: it must account for the fact that rules can influence behavior and are not just devices to describe it.

The obvious way to combine these insights into a unified theory is to stipulate that institutions are rules in equilibrium, and the rules are summarized using some form of symbolic representation. This idea has been around for a while, in the attempt to combine the best aspects of rule-based and equilibrium-based theories of institutions. According to Avner Greif and Christopher Kingston, for example,

> Despite their differences, the institutions-as-rules and institutions-as-equilibria approaches have much in common and are best viewed as complements rather than substitutes. . . . The role of "rules," like that of other social constructs, is to coordinate behavior. Because there are multiple potentially self-enforcing expectations in a given situation, coordination mechanisms, including rules, play an essential role in generating regularities of behavior and social order. Rules fulfill this coordinating role by specifying patterns of expected behavior, and also by defining the cognitive categories—signs, symbols, and concepts—on which people condition their behavior. (2011: 15, 28)

Greif and Kingston do not offer a precise definition of institutions, but their conception fits some accounts that are on offer. Masahiko Aoki for example proposes the following definition:

> An institution is a self-sustaining, salient pattern of social interaction, as represented by meaningful rules that every agent knows, and incorporated as agents' shared beliefs about the ways the game is to be played. (Aoki 2007: 6)

The rules are symbolic markers that represent equilibria (or parts of equilibria) and help the players use a particular coordination device. Unlike in rules-based theories, the concept of pattern (equilibrium) is central in this theory. But unlike "pure" equilibrium-based theories, this account brings at center stage the representation of the equilibrium by means of symbolic markers (rules). This way, we obtain a satisfactory, consistent, and empirically adequate conception of institutions.

REFERENCES AND FURTHER READINGS

There are many equilibrium-based theories in the social science literature, stemming from Lewis's (1969) seminal work on conventions. See, for example, Schotter (1981), Calvert (1998), Greif (2006), Binmore (2010). The account presented here is close in spirit to Aoki's (2001, 2011) and Greif and Kingston's (2011) "hybrid" theories. The use of the hawk-dove game to represent animal and human conflicts over contested resources goes back to Maynard Smith's (1982) work on evolutionary game theory. See also Sugden (1986) and more recently Gintis (2007). On correlated equilibria see Aumann (1974, 1987), as well as Vanderschraaf (1995, 1998, 2001) and Gintis (2007, 2009). The theory of institutions as rules in equilibrium has been discussed in a symposium recently published in the *Journal of Institutional Economics.* The commentaries of Aoki (2015), Binmore (2015), Sugden (2015), and Smith (2015), in particular, address some of the issues discussed in this chapter. Hindriks and Guala (2015b) reply.

CHAPTER 5
CONSTITUTION

Few philosophers have had a greater impact on social ontology than John Searle. *The Construction of Social Reality* (1995) articulates in detail a theory that Searle had sketched already three decades earlier, in his seminal work on speech acts. Searle's theory falls squarely in the rules camp and—unsurprisingly, for a philosopher of language—places language at center stage.

The key idea is familiar: institutions are systems of rules. Searle's rules however are different from the "if X then do Y" statements that we have encountered so far. They are *constitutive* rules, that is, statements that define what a certain entity or class of entities is, what social function it has in a given context, and the conditions it must satisfy in order to have that function. Searle has proposed a memorable grammar of social reality, a formula that captures the underlying logic of all institutional facts. He has claimed that "an institution is any system of constitutive rules of the form *X counts as Y in C*" (2005: 10).

In his writings Searle never makes use of the theories and concepts of social science. On the contrary, he often expresses contempt for a body of doctrines that—in his view—has failed to explain the deep structure of social reality. In his writings he claims that his approach is radically different from, and incompatible with, the approach of economists and social scientists who study institutions. In this chapter I will argue that this is not the case. The main line of argument, developed with Frank Hindriks, aims at showing that Searle's approach not only is compatible with the approach of modern social science, but can be derived from the rules-in-equilibrium theory (with a little help from theoretical terms, as I will explain shortly). This means that from an ontological point of view Searle's theory is equivalent to a theory that makes use of only the "if X then do Y" statements that constitute the bulk of the rules-in-equilibrium approach.

First, however, it will be necessary to illustrate Searle's view briefly. The notion of constitutive rule is usually traced back to a distinction originally made by John Rawls (1955). Rawls distinguished between a "summary view" and a "practice conception" of rules. A "summary" rule describes a set of particular actions that are in some respect similar, while a "practice" rule defines a certain type of actions *before* they are instantiated in specific cases. Rawls mentions the game of baseball as an example: some rules (the rules of the game of baseball) seem to be conceptually prior to the specific actions that take place on the pitch. Trying to "steal a base" or to "hit a strike" would not make sense in the absence of more fundamental rules that define the nature of the practice that we call "baseball."

> No matter what a person did, what he did would not be described as stealing a base or striking out or drawing a walk unless he could also be described as playing baseball, and for him to be doing this presupposes the rule-like practice which constitutes the game. (Rawls 1955: 25)

The distinction proposed by Searle mirrors Rawls's distinction closely. While *regulative* rules describe or guide the behavior of individuals engaged in certain social interactions (in analogy with "summary" rules), *constitutive* rules are necessary to define or create new social entities, roles, and facts (as in the "practice conception").

Clearly the rules that we have seen so far are regulative rules that prescribe specific actions in specific games. They take the form of "if X then do Y" statements. A constitutive rule in contrast stipulates that

X counts as Y in C,

where X is a preinstitutional entity (like sending the ball out of the pitch), Y is a "status function" (like hitting a home run), and C stands for a domain of application (during a game of baseball). The formula is simple and memorable, and has had huge success in the philosophical literature. It does, however, hide some complications that are worth unveiling.

Searle's primary goal with the X-counts-as-Y formula is to explicate the relation between noninstitutional reality—the realm of physical entities, for example—and institutional facts. X terms in the formula refer to noninstitutional or preinstitutional facts, and Y terms refer to institutional facts (events, entities, properties). Searle uses the expression "status function" to express the idea that the creation of institutions implies the assignment of new functions to objects or events that otherwise would not have that function. Another way to put it is that institutional reality is "built upon" or "superimposed" on the natural world.

Let us call X the *basis* of an institutional fact. X may be a token entity, like a stone for example, when a group of children says "this stone counts as the pitcher's mound in our game of baseball." But in many cases constitutive rules apply to entire classes or types of entities. The Italian Constitution for instance states that the Members of Parliament elect the President of the Republic (article 83), and that any citizen who is at least fifty years old and who enjoys full political and civil rights may be elected President of the Republic (article 84). In Searlean terms, these two principles may be restated as follows:

> Any citizen who is at least fifty years old, who enjoys full civil and political rights, and who has been elected by

> the members of Parliament (X) counts as President of the Republic (Y) in Italy (in C).

Two things are worth noting: first, the expression "counts as" is meant to convey the conventional character of the X-Y relation. An equivalent formulation could be "X is considered as Y by the members of a certain community." But we might as well use a compact notation and simply say that "X is Y" in the community. Second, the X entities must satisfy certain conditions in order to be Y. The reason is quite obvious: while in the case of a token entity the basis can be identified simply by pointing to the thing in question ("*this* is the pitcher's mound"), such a procedure cannot work when X refers to a type of entity. The constitutive rule must give some indication of the sort of things that count as Y, by stating the properties or conditions that the tokens must satisfy in order to acquire the institutional status. Notice that some conditions (like age, for example) are physical or biological, but others (civil rights) are not. The important point is that they are preinstitutional with respect to the institutional role that is at stake (President of the Republic in this case).

The conditions that X must satisfy in order to count as Y are hidden in Searle's formula. But they can be made explicit by stating the properties P that identify exactly which entities of type X are Y. If an entity does not have these properties then it does not count as Y. Two notations are possible: we can spell out the full formula

if P then X is Y in C,

or instead use the C-term to refer to the properties or conditions that must be satisfied for X to count as Y:

X is Y if C.

The second solution is less faithful to Searle's original notation, but since it is simpler and preserves the beautiful musical rhythm of the formula, I will adopt it from now on.

Let us call the set of properties that X must have in order to count as Y, the *conditions of satisfaction* for Y. In what follows I will take C to denote the conditions of satisfaction, unless otherwise specified. For instance, if he is touched by a pitched ball outside of the strike zone, and he attempted to avoid it (or had no opportunity to avoid it), and he did not swing at the pitch, then a batter counts as a base runner.

Of course a single rule does not exhaust a complex institution like baseball. A typical institution involves a large number of constitutive rules. As Searle rightly points out, the rules act as constraints on the actions of the individuals who are involved, but at the same time also create opportunities that otherwise would not exist. You cannot hit a home run or steal a base if the institution of baseball does not exist, and you cannot impeach the President unless this procedure is codified in the constitution (if you try to do something similar regardless of the constitution, it would be a coup d'état, not an impeachment). Most of the things that take place in parliament or on the baseball diamond are *institutional facts*, in the sense that they would not be what they are if the appropriate constitutive rules were not in place. Thus "institutional facts only exist within systems of constitutive rules" (Searle 1995: 28).

The distinction between constitutive and regulative rules supports one of Searle's most important claims, namely that institutional reality presupposes the existence of language: if institutions make new activities and new facts possible, and institutions are systems of constitutive rules, then the capacity to formulate such rules is a prerequisite of institutional reality. The latter comes into existence in virtue of the rules: institutions presuppose language.

But what kind of rules are needed? Are constitutive rules necessary, or do regulative rules suffice for the creation of institutions? In the rest of this chapter I will ask whether the distinction between regulative and constitutive rules is genuine. Because if it was not, the consequences for Searle's overall theory would be quite serious. Fortunately I will not explore an unchartered territory. In a series of recent papers Frank Hindriks has shown that constitutive rules can be decomposed in simpler elements, and some of these elements look worryingly similar to regulative rules. What is worse, Hindriks has argued that the terms of the decomposed formula that are *not* regulative rules are eliminable. But then, if only the regulative rules do substantial work, the key distinction upon which Searle's theory is based does not hold.

To appreciate this point, we shall return briefly to the Sobat Valley, where the Nuer and the Dinka have been grazing their cattle peacefully. Recall that in the original story the Nuer and the Dinka had found a convenient solution to the problem of grazing. Using the river's bed as a correlation device, they had assigned the land that lies north of the river to the Nuer, and the land that lies south to the Dinka. Now whenever the Dinka and the Nuer look at a patch of land, they immediately notice its location and apply their conditional strategies—they play their part in the following correlated equilibrium:

[R] (i) Graze if the land is north, do not graze if it is south of the river (G if N, NG if S).
(ii) Graze if the land is south, do not graze if it is north of the river (G if S, NG if N).

These statements are regulative rules in Searle's terminology. They describe preexisting activities, patterns of behavior that have evolved in the past for one reason or another. It is worth stressing that the members of the two tribes need not be aware of the reason why they are following these rules. To strengthen the equilibrium, they could invent a magical theory that gives them extra reasons not to cross the river's bed ("there are evil spirits on the other side!"). Alternatively, the equilibrium may be sustained simply by imitation: the youngsters may behave as the elders did, and since the elders have always stopped at the river's banks, the youngsters keep doing the same. The existence of conventions does not require a deep understanding of the rules that define the correlated equilibrium of the game.

So far nothing that looks remotely similar to a constitutive rule plays a role in our story. The reason is that institutional terms like "border," "territory," or "property" do not appear in the vocabulary of the Dinka and the Nuer. The equilibrium in the Sobat Valley requires only that the two tribes follow the regulative rules [R]. And yet, intuitively, the social arrangement that has emerged in the Sobat Valley looks very similar to an institution. An anthropologist who has just entered the valley and noticed the grazing patterns of the two tribes would conclude that an informal border divides the territories of the two tribes. But borders and territories are institutional entities, and there cannot be institutions without constitutive rules, according to Searle. So either the anthropologist is wrong, or constitutive rules are redundant—regulative rules like [R] are sufficient for the existence of institutions.

The issue is partly verbal, of course. One may simply decide to restrict the notion of institution so as to satisfy Searle's criterion. Conventions would be excluded by definition, because they involve only regulative rules. And this may be a fair price to pay for a philosophically founded taxonomy of social entities. This move, however, is bound to fail. It is possible to show, in fact, that constitutive rules may be derived from regulative rules at no cost. Here is one version of Hindriks's argument.

The terms that appear in [R] refer to specific actions ("graze on the north side," "do not graze on the south side") that apply to the Dinka-Nuer game. But nothing prevents one from inventing a more general concept that implies grazing as one of its consequences. Let us introduce the term "property," to denote all the patches of land where the members of a tribe graze their cattle. Since this is a somewhat artificial concept, much simpler than our full-blown notion of private property, let us add an asterisk to mark its peculiarity. A (partial) characterization of the term property* is given by a *status rule* like this:

[S] If a piece of land is Nuer's property*, then the Nuer graze it.

The status rule introduces the new concept of property*, and relates it with previously understood terms like land, grazing, and the Nuer people. Another way to put it is that [S] partially explicates the meaning of the term "property*." The use of a new term requires the specification of some criteria of application however. When is it appropriate to say that something is Nuer's property*, and when is it not? This is determined by a *base rule* like this:

[B] If a piece of land lies north of the river, then it is Nuer's property*.

And here comes the crucial step of the argument: the conjunction of a status and a base rule forms a constitutive rule in Searle's sense (B + S = CR):

[CR] If a piece of land lies north of the river, then it is Nuer's property*, and if a piece of land is Nuer's property*, then the Nuer graze it.

The grammatical form of [CR] is

if C then X is Y, and if Y then Z.

Note that the term Z did not appear in Searle's original formula. In the new formula (let us call it the XYZ formula) Z is used to refer to a set of actions that are associated with the term Y. In our case Z is grazing or not grazing, for example. The new formula has the advantage of clarifying that Searle's distinction between regulative and constitutive rules does not reflect a substantial distinction between types of social entities. Introducing the term property* does not add anything new, given the definition in [S]. Property* existed before the constitutive rule was formulated explicitly. The rule introduced a new term only to refer to behaviors described by the regulative rules, but the term is eliminable. If we eliminate the new term from CR we obtain

> [CR'] If a piece of land lies north of the river, then the Nuer graze it,

or, in more abstract terms,

if C then Z.

This demonstrates that institutions like property* may be explicated using rules of the standard if-then format that regulate preexisting activities. Or, to put it differently, regulative rules can be extended to create constitutive rules at no cost, merely by inventing new Y-terms, because the Y-terms refer to regulative rules. This means that institutional terms in principle are eliminable because the CR rule creates nothing that did not exist before or that could not exist without it.

That institutional terms are elimin*able*, of course, does not mean that they must be elimin*ated*. On the contrary, the introduction of theoretical terms has an important pragmatic function: it promotes economy of thought and language, bundling together a set of regulative rules that we use to coordinate behavior in a set of related games. The XYZ formulation, however, reminds us that the institutional entities denoted by the Y-terms do not exist alongside or on top of regulative rules in equilibrium. To say the latter would be to commit a fallacy of double counting, analogous to adding a chair to the list of atoms that are located just under my bottom. These entities must not be counted twice because they are one and the same thing.

Two objections must be faced at this point. First, one may point out that the CR formula merely plugs some regulative rules into the original formula. The base rule [B] has exactly the same grammatical form of Searle's XY formula (if C then X is Y). So why should we invent a new name for it, and use the "constitutive rule" label instead for the augmented XYZ formula? The reason is that Searle's version of the constitutive rule formula is elliptic. Unless we say what Y stands for, it does not mean very much. The formula must be made explicit by specifying the content of Y. Once this has been done, we can see that the full constitutive rule has the XYZ grammatical form. This is a *substantial*, not just a grammatical point, and for this reason it is helpful to use the [CR] label for the full formula. The point is that without a Z-term the constitutive rule cannot affect people's behavior in any way.

Still, it is not clear that this is the right way to spell out the content of the original formula. Perhaps there are other ways to do it, perhaps there is more to Y than merely a set of regulative rules. This is a fair comment, and for the time being I can only ask to be patient. The only claim so far is that if we introduce institutional terms like property* to refer to the Dinka-Nuer convention, we must keep in mind that they refer to regulative rules. But nothing guarantees that the same holds for more complex institutions like real private property. For the time being let me just say that the XYZ approach looks promising: although we have not satisfactorily accounted for the real thing yet, at least we have analyzed something that looks remarkably similar to it.

Before I proceed, a few remarks on the function of Y-terms are in order. As Hindriks points out, a constitutive rule is "a theory of sorts." Its theoretical terms refer to phenomena and entities that exist independently of the theory itself. One reason why it is useful to have property* in one's vocabulary is that it spares the trouble of listing the actions that constitute the equilibrium between the Nuer and the Dinka. But the theoretical term property* can in principle be dispensed with, by formulating the theory entirely in regulative language. In fact this happens with lots of institutions that we never bothered assigning a name. We do not have a name for the institution of driving on the left-hand side of the road, for example. We could call British and Japanese drivers "leftists," if we wanted to, but we never felt the need to introduce a theoretical term in our language for this type of convention.

Why is that? A plausible explanation is that such an institution is too simple to deserve a name. The institution of driving on the left is exhausted by a single regulative rule (if you are in Japan, then drive on the left). Theoretical terms are useful if they promote economy of language and thought, and in fact we usually give names to institutions—like property or marriage—that are more complicated than driving on the left.

A single institutional term is usually introduced to refer at once to a bundle of rules that govern behavior in a various related strategic interactions. The bundle in other words is far from arbitrary or "gerrymandered." Across different societies, for example, the rules of marriage regulate activities like the rearing of children, insurance, inheritance, and sexual access. But we do not use a single term for all these rules by coincidence: on the contrary, the rules that regulate behavior in one domain are functional to promote specific behaviors in other domains. Rearing children, for example, requires considerable investment in terms of energy, emotions, time, and money. So the rules of child-rearing, the rules that regulate sexual behavior, and those that regulate economic transactions are partly interdependent.

This does not mean that the bundles are static, or that the various rules are so tightly connected that they cannot be changed in a piecemeal manner. Institutions are evolving entities, that are constantly revised and adjusted depending on the circumstances. Recall also that most institutions are constituted by informal rules, at least in part. This provides flexibility, because informal rules are more easily renegotiated than formal ones, but it also introduces a certain degree of vagueness regarding what a given institution is at any point in time. It is an informal rule of marriage in many countries that women must sacrifice their careers for rearing their children, for instance. But the informal rule in some countries is faltering, to the extent that it has become unclear what is the exact composition of the bundle of rules that constitute marriage. The meaning of "marriage" may be partly indeterminate, and different groups may interpret the concept in different ways. (I will return to this issue at the end of the book.)

Another important feature of real institutions is that they involve rights and obligations. Economists for example say that property is a "bundle of rights"; marriage similarly involves rights and obligations toward one's spouse and children. Rights and obligations introduce a new element—normative power or deontic force—that is absent in the simple institution of property* analyzed so far. As we shall see, the main point of charging institutions with normative force is to make certain behaviors more costly, and thus modify the equilibria that are available in a class of games. Since this normative aspect of institutions plays an important role in the theories of Searle and other philosophers, it will be necessary to discuss it in detail in the next chapter.

REFERENCES AND FURTHER READINGS

Wittgenstein's discussion of language games in the *Philosophical Investigations* (1953) inspired various distinctions that are broadly similar to the one between regulative and constitutive rules. Besides Rawls (1955), see also Midgley (1959). Morin (2011) presents a series of critiques to the regulative-constitutive rules distinction. The essential elements of the constitutive rules approach are already in Searle's *Speech Acts* (1969), but the theory is developed fully in *The Construction of Social Reality* (1995) and *Making the Social World* (2010). Searle, unlike Lewis, treats language as the most fundamental institution upon which the whole structure of social reality is based. Other relevant statements of the theory can be found in Searle (2005, 2009). For the explicit inclusion of conditions of satisfaction in the "X counts as Y" formula, see in particular Searle (2010: 99). Ransdell (1971) and Conte (1988) have pointed out that in order to make the meaning of constitutive rules explicit it is necessary to expand the formula by adding another term that specifies the import of the institutional term. Hindriks developed this idea in his doctoral dissertation (2005) with the aim of unifying the constitutive rules approach with Lewis's theory of conventions (see also Hindriks 2009). The connection with correlated equilibria is explained in Guala and Hindriks (2015) and Hindriks and Guala (2015a). Searle's opinion on our attempted unification can be found in his contribution to the *Journal of Institutional Economics* symposium (Searle 2015), to which we reply in Hindriks and Guala (2015b). Another recent article that addresses the relationship between game theory and constitutive rules, from a slightly different perspective, is Hédoin (2015). Finally, I must confess that I have no idea how the game of baseball works. If the rules cited in the text are wrong, it is Wikipedia's fault.

CHAPTER 6
NORMATIVITY

The argument presented so far is meant to show that the constitutive rules theory can be translated without loss in the language of the rules-in-equilibrium theory. The previous chapter has laid the foundations for the translation, focusing on the Nuer-Dinka game as a paradigmatic example. The use of such a simple model, however, raises some concerns that must be dispelled before we can accept the equivalence thesis. To begin with, it is apparent that the translation relies heavily on redescription. The reconstruction of the XY formula in terms of regulative rules (the XYZ formula) works well enough for an institution like property*. But property* is not real private property. It is a toy version of the institution of private property, much simpler than its real-world counterpart. And this prompts some questions: Can all the features of real institutions be explained in terms of regulative rules? Is the translation going to work? Can we really make do without theoretical terms, and hence constitutive rules, at no cost? Or are we going to miss some important element of institutions along the way? This chapter addresses these questions, focusing on two points in particular: the *functional properties* of institutions and their *normative power.*

It is a cornerstone of Searle's theory that constitutive rules (X is Y if C) assign new status functions to objects.

> The Y term has to assign a new status that the object does not already have just in virtue of satisfying the X term; and there has to be collective agreement, or at least acceptance, both in the imposition of that status on the stuff referred to by the X term and about the function that goes with that status. (Searle 1995: 44)

The functions associated with the Y terms are "new" in the sense that the preinstitutional objects (X) cannot fulfill them in virtue of their preinstitutional properties. Consider money: a paper bill counts as money if it is issued by the Central Bank. To say that X is money is to say that it performs the typical functions of money—it works as a medium of exchange, store of value, and unit of accounting. But the physical properties of a paper bill are insufficient by themselves to ensure that the bill does all these things, since there are physically identical paper bills that do not fulfill these functions (they are counterfeit bills, or currency that is no longer in circulation, for example). Moreover these properties are unnecessary, because other things are money even though they have very different physical shapes (like shells, coins, and electronic bytes).

So, how are the functional properties of institutions accounted for by a theory that considers the Y terms inessential and in principle eliminable? According to Hindriks's view, the Y terms are theoretical terms that we use to name institutions, that is, the bundles of regulative rules that govern our interactions in sets of related games. So if the Y terms refer to functions, the functions must somehow be fulfilled by the regulative rules. Since the notion of function is strictly related to purpose and goal, the question can be rephrased like this: What is the purpose of regulative rules? What goal do they help to accomplish?

The answer is straightforward: regulative rules, like the rules of traffic, the rules of private property, or the rules of money, facilitate coordination. Achieving coordination—with low effort and high reliability—is the main function of institutions.

	Stop if red, go if green	Go if red, stop if green
Stop if red, go if green	1, 1	0, 0
Go if red, stop if green	0, 0	1, 1

FIGURE 6.1. The traffic lights game.

There is a strict relation between coordination equilibria and functions. Recall that an equilibrium is a profile of actions, and the actions can be formulated as rules. So whenever a combination of actions leads to an outcome that is better than a relevant alternative, it is natural to say that the rules that indicate the actions have a function—the function of making us better off. (I am speaking of "function" here without assuming that the goal is intended or represented in anybody's mind; this is similar to the way in which biologists use the term, although we do not need to assume that the rules are adaptive or have survived a process of natural selection.)

Consider the traffic lights game of Figure 6.1. The rule "stop if the light is red, go if it's green" makes us better off, making the traffic flow smoothly and preventing accidents. The relevant alternative in this case is lack of coordination (in the bottom-left and top-right corners of the matrix). The same holds for any coordination game, by definition, because in a coordination game any equilibrium is better than the relevant out-of-equilibrium outcomes. Since institutions are conditional strategies of coordination games, and the strategies are regulative rules, the regulative rules have a function. Equilibrium profiles in coordination games go hand in hand with the fulfillment of functions.

One may point out that in Searle's theory the function is assigned to an object or a state of affairs (X), while in the unified theory it is associated with a set of rules. But this is a superficial difference: because the rules have a function, the coordination device (the traffic light) acquires a function too. This is the kernel of truth in Searle's formula. Clearly a pair of alternating red and green lights could not have any function by themselves. They do acquire a function, however, when they are used to correlate our behavior in a coordination game. But to say that the creation of institutions involves the assignment of functions to a physical object is elliptic and obscures the most important part of the mechanism. The creation of institutions involves first and foremost the creation and the implementation of *rules* that make us better off, compared to an alternative outcome. If the rules condition our behavior on external events, then the events (or correlation devices) help to fulfill that function.

The fact that the Y term is eliminable, then, does not do any harm. Functions continue to be fulfilled by institutions whether or not they are named explicitly by means of a theoretical term. And a theory based on regulative rules—like the unified theory—can account perfectly well for the functional properties of institutions.

The second important feature that seems to be missing from the proto-institution of property* is *normative power*. In his writings Searle draws a tight connection between the functional role of institutions and their deontic features:

> Because the creation of institutional facts is a matter of imposing a status and with it a function, in general the creation of status-functions is a matter of conferring some new *power* . . . . In the simplest case, the Y term names a power that the X term does not have solely in virtue of its X structure. (Searle 1995: 95)

> The simplest test for whether a phenomenon or fact is genuinely institutional is to ask, Does its existence imply deontic powers, powers such as those of rights, duties, obligations, requirements, and authorizations? (Searle 2010: 91)

"Deontic" is a technical term derived from the Greek word *deon* ("obligation," "necessity"), that philosophers use to refer to normativity. Searle claims that the introduction of Y terms has the effect of creating new normative powers that did not exist before. Several commentators have noticed that it is not at all clear how deontic properties can be derived from the notion of function. It seems entirely possible that something (a rule for example) has a function without there being any normative reason to conform with it, apart from purely instrumental reasons. Wearing black at funerals for example may have the function of communicating one's sorrow, but if we do not want to communicate our sorrow then it seems that we have no obligation to wear black in spite of what the institutional rules say.

Some philosophers claim that there is an intrinsic normative aspect in every rule, for the mere fact that it is a rule. This claim usually boils down to the observation that when a statement is used to issue a directive ("wear black at funerals!"), a set of actions is sorted in two categories—those that conform with the rule and those that do not. Interpreting a rule thus requires the definition of a standard, a criterion to distinguish correct from incorrect applications of the rule. But this trivial sense of normativity is not what is at issue: the question of the deontic power of institutions is the question of whether one should conform with the rule in the first place. Using the same example again: I may agree that wearing jeans constitutes a breach of the traditional dress code for funerals, but at the same time I may well believe that there is no reason why I should not wear jeans in such occasions. To say that a rule can be transgressed and to say that a rule ought to be followed are completely different matters.

Having cleared the ground from these preliminary issues, the question we have to address in this chapter is whether by making Y terms dispensable the unified theory also dispenses with the deontic power that is often associated with such terms. Even if we cannot demonstrate that institutional terms and deontic powers are conceptually linked, in fact, from an empirical point of view there is no doubt that they often are correlated. Real institutions like private property involve the attribution of rights and duties, for example, and the very language that we use is proof of that. A piece of land may be Nuer's property even though the Nuer do not *actually* graze it. The point of it being their property is that the Nuer *can* graze it if they want. Whether they do it or not depends on various contingent circumstances. For example, the Nuer may decide that they prefer to let the grass grow, or plough the land, or rent it out to an oil company. They are permitted to do all these things, if it is their property, while the Dinka are not.

To simplify, and to refer to all the things the Nuer can do with their land, it is convenient to adopt a generic term like "use." The equilibrium in the Sobat Valley then should be formulated by means of normative rules like the following:

[R'] If a piece of land lies North of the river, then the Nuer can use it.

This statement is expressed in deontic terms, using the modal operator for *permission* (can, may, is allowed to). Permission is one of three basic operators of deontic logic. The other two are *obligation* (ought, must, is obliged to) and *prohibition* (cannot, may not, is forbidden). The three operators are mutually interdefinable, that is, one can express any deontic statement using one of the other operators plus negation. For example:

You *must* do A = you cannot do not-A.

You *can* do A = you do not have to do not-A.

This is not a pure grammatical or logical curiosity: it means that any institutional rule expressed in deontic form imposes or lifts some constraints on behavior. Moreover, it is an interesting fact of institutional rules that they usually create *mutual* constraints. My rights are your obligations, and vice versa.

If we say that "students may appeal against a mark assigned by a lecturer," for instance, we mean that they do not have to abstain from filing an appeal. This in turn means that no one (especially a university lecturer) is allowed to block a student's appeal. In our previous example, [R'] in fact is equivalent to [R'']:

> [R''] If a piece of land lies North of the river, then the Dinka *cannot* use it.

That Jack can do X means that no one can force Jack to do not-X. For simplicity, when a convention is augmented with deontic force imposing extra constraints on players' behavior, we shall call it a *norm*. But what is a norm exactly, and where does deontic power come from? When philosophers claim that a rule has normative force they generally mean that people have reason to follow it. The notion of reason is very broad, however, and intuitively we should try to be a little more specific. In this vein Searle claims that deontic rules provide a special type of reason to act:

> Deontic powers have a unique trait, . . . I think, uncommon and perhaps unknown in the animal kingdom: once recognized they provide us with reasons for acting that are independent of our inclinations and desires. (Searle 2010: 9)

This specific view is controversial however. According to a tradition that goes back to David Hume, all human actions—including actions that are performed to comply with norms—are driven by desires. The controversy between Humeans and their opponents is one of the central unresolved issues in the philosophy of action, and although it raises deep and important philosophical questions it would be a mistake to make the issue of normativity depend on it. That the problem of the nature of institutions should depend on whether human action is always desire-driven or not sounds rather odd. Instead of trying to adjudicate the issue, then, it is preferable to work with a definition of normative power that is neutral regarding the Humean thesis.

To be as general as possible, let us say that norms provide reasons for acting that are independent of *some* desires. It will not be necessary to specify which desires exactly. As a matter of fact, most moral and social norms prescribe behaviors that are independent or even in conflict with the satisfaction of our personal needs and wants—our *self-interested desires*, we might say, provided we care to interpret self-interest broadly, to include for example the welfare of our relatives and friends. But in principle there might be norms that go against other types of desires as well (although I suspect that they are quite rare).

This formulation is compatible with the Humean view: someone might happen to have a desire to abide by a rule. The point, in any case, is that she still has reason to do so in the absence of such a desire. Furthermore, she also has a reason to follow the rule if she has a conflicting desire. The next question is: how can this reason be incorporated in the rules-in-equilibrium theory? Recall that in game theory individual motives and goals are summarily represented by the structure of payoffs. This representation seems to imply a commitment to consequentialism—the view that people care only about the outcomes of their actions, as opposed to the way in which the outcomes are attained—and for this reason some philosophers dislike it. But provided one is not strict about the motives that are incorporated in the payoff function, this modeling strategy allows the representation of deontic powers within an equilibrium framework.

Notice that by following this strategy the unified theory provides only a formal apparatus to represent normativity, but it remains neutral about the nature of normativity or where it comes from. And I think that this is just as it should be. Normativity is one of the thorniest issues in contemporary philosophy, and it would be foolish to make a theory of institutions depend on a specific account. Some philosophers and social scientists believe that normativity can be analyzed in terms of mutual expectations and the feeling of resentment that we experience when our expectations are frustrated; others believe that normativity requires a stronger notion of collective agreement or joint intention; some philosophers and social scientists argue that normativity depends on emotions; and still others believe that normativity has to do with the possibility of justifying our actions by means of rational arguments.

Whether any of these accounts is able to explain normativity in a satisfactory way is an open issue that I do not want to settle here. In fact choosing among them may not be very sensible: if normativity is important for institutions, then it is likely to take several different forms. As an analogy consider the multiple ways in which an organism tries to accomplish a goal that is important for its survival. If perceiving the existence of a prey is important, it is likely that a predator has more than one way to accomplish the task (by vision, hearing, and smell, for example). Similarly, normativity probably has different sources and many facets, which means that more than one account is likely to be right.

So instead of asking what normativity is, I will ask what it does, or what its function is. Although it is in line with the broadly functionalist conception of institutions that runs through the book, this strategy will inevitably leave some readers unhappy. One reason is that this approach does not allow one to make any substantial normative assessment of institutions. It does not allow one to distinguish good from bad institutions—dictatorship from democracy, capitalism from socialism, or polygamy from monogamy, for example. My own view is that judgments of this sort belong to the realm of ethics, rather than social ontology, and that it is best to keep the two projects separate. Other philosophers disagree and have tried to construct more robust theories of institutions—with mixed results in my view—but I will let the readers make up their own minds on this point (there are references at the end of the chapter).

Let us return to the main topic now. How can normativity be represented in the framework of the unified theory? A convenient way of modeling the constraints introduced by norms is in terms of costs. As long as the latter are understood in a sufficiently broad way, we can say that compliance with a norm carries costs while conforming to a nonnormative rule or convention (a rule of thumb for example) does not. For example, a littering norm imposes the cost of looking for garbage bins. A norm against cheating imposes the opportunity cost of not having extramarital affairs, and so forth. Such costs call for compensation (counterreasons) in order to be overridden. And counterreasons may be represented as negative incentives—other costs, effectively—that deter norm violations.

The representation in terms of costs is compatible with various enforcement mechanisms. Social scientists distinguish between internal and external costs, corresponding roughly to a distinction between "internalized norms" and "externally sanctioned norms." The first type of norm has been central in the sociological tradition, dating back to the work of Talcott Parsons: social norms according to this interpretation are internalized prescriptions that people follow partly out of habit, and partly because they do not want to feel bad about it. The costs may be seen as a form of self-inflicted punishment, in the form of guilt feelings. Another equally influential tradition sees social norms as supported by external punishment instead. In this case, the costs are inflicted by other members of society, either professional enforcers (judges, policemen) or fellow citizens who just happen to witness the transgression. Gossip, blame, reproach are all standard enforcement devices in communities governed by informal norms. Unwritten rules may have the force of a law, if the sanction is generally known and expected.

Elizabeth Crawford and Elinor Ostrom (1995) introduce a delta parameter to capture the costs involved in violating a norm. In the case of externally sanctioned norms, the cost of transgression is generally known to the members of the community, and can often be quantified objectively. A century ago, for example, the cost of refusing to marry a Sicilian woman after a one-night stand was equal to the expected value of death. The cost of littering the floor in Singapore is three hundred dollars weighed by the probability of being fined. When the costs are internalized, deltas are more difficult to measure of course. But measurement is not impossible, if we have enough information about the desires that the norm is supposed to trump. The key is to know what people would have preferred to do if the norm had not been in place. The abiding citizen who regularly clears the pavement in front of her house from the snow invests precious time that she might have liked to spend differently (by watching TV in her warm living room, for example) had she not internalized the norm.

Representing normative power by means of delta parameters (costs) facilitates the extension of the unified theory beyond the realm of coordination games. Many social theorists have pointed out that institutions improve the performance of players in games where there is an individual incentive to deviate from the socially optimal rule. A classic case is the prisoner's dilemma game (Figure 6.2). Instead of multiple equilibria—as in the coordination games examined so far—there is only one equilibrium in the one-shot prisoner's dilemma (DD). Augmenting the game by means of conditional strategies does not help, so there is no way to solve a prisoner's dilemma using an external correlation device. Defecting strictly dominates rule following. By means of a suitably large delta, however, one can transform the dilemma into a game with two equilibria, CC and DD. Normative rules thus can turn a cooperation problem into a coordination problem: norms change the games that people play. And, of course, they also work as coordination devices (choreographers) in the coordination games that have just come about.

Suppose there is a rule in the population that says "if the other player cooperates, then you ought to cooperate, otherwise defect," and the rule has normative force. This means that an extra cost (delta) must be subtracted from the payoffs, as in Figure 6.3. The payoffs in these two games may be interpreted

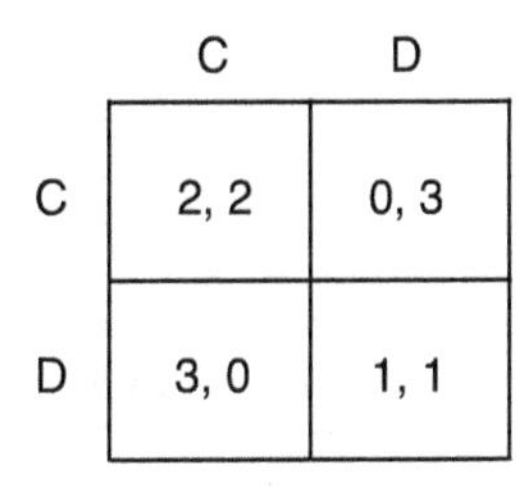

FIGURE 6.2. The prisoner's dilemma.

	C	D
C	2, 2	0, 3–δ
D	3–δ, 0	1, 1

FIGURE 6.3. Prisoner's dilemma modified by delta parameters.

in various ways: one possibility is to take the numbers in the original game (Figure 6.2), as representations of prenormative desires. In the new game (Figure 6.3), then, the numbers may represent individual goals after the internalized desire to comply with the norm has been taken into account. Alternatively, the modified payoffs of Figure 6.3 may incorporate new information that has become available to the players, for example when they discover that a system of punishments has been set up to deal with those who transgress the rule. In any case the deltas represent the force of normative rules, taking a given (prenormative) game as a benchmark.

Depending on the force of the rule, the second game may turn out to be quite different from the first one. If delta is at least as large as one unit of payoff, then the prisoner's dilemma is transformed into another simple game, where DD and CC are both equilibria. If delta is equal to three, for example, we obtain the hi-lo game of Figure 6.4.

Notice that norms then "solve" dilemmas of cooperation only in a peculiar sense. Games such as the one-shot prisoner's dilemma cannot be properly solved, because there is no way to

	C	D
C	2, 2	0, 0
D	0, 0	1, 1

FIGURE 6.4. Modified prisoner's dilemma with delta equal to three.

escape the disturbing conclusion that players ought to defect, without changing the rules of the game. The only "solution" is to change the game itself, and this is precisely what institutions endowed with normative power can do. They create new equilibria introducing costs that make defection unattractive, at least within a certain range of payoffs.

But this is not the only function of norms. Deontic powers are useful in games of coordination as well. This may sound strange, since the players do not have any reason to deviate from a coordination equilibrium. Adding deontic power seems gratuitous, if there is no conflicting desire to counteract. Deontic power however is unnecessary only if the players have full information and never make mistakes. In real life people have the unfortunate propensity to make mistakes, to misinterpret the signals sent by the choreographer, and sometimes they are also uncertain about the payoffs of the game. In such cases it is useful to have some extra mechanism that helps enforce conformity with a rule. People pay more attention to the signals they receive and to the structure of the game, if they know that they will pay an extra cost in case of noncompliance. Upon reflection, it is not surprising that many conventions have a tendency to turn into norms.

Norms thus help fulfilling two key functions of institutions, highlighted by North (1990): they stabilize behavior and make it more predictable in situations of uncertainty; but as pointed out by Searle they also create behaviors that did not exist before, by changing the payoffs of a game. Thus, norms help explain not only the *persistence* of institutions, or how they can remain in force in the face of incentives to deviate; sometimes they also explain the *emergence* of new equilibria, when they are introduced ex novo. A central authority, like a government or a recognized leader, can reshape a game with "bad" equilibria, such as a prisoner's dilemma, and turn it into a better game, such as hi-lo. New rules may be introduced by decree: if the elders meet in the middle of the plain, and declare that from now on the Dinka ought to graze south and the Nuer ought to graze north of a salient location; and if the decree is supported by credible formal and informal sanctions, then the players will recognize that the game has changed and that a new equilibrium has been created. The punishment mechanisms will make the new equilibrium more salient than it would have been in virtue of the decree alone. So the function of normativity can be accounted for within the same framework that we have used to explain the convergence on equilibria in coordination games.

It is important to stress that the rules that change behaviors and create new regularities are *regulative* rules though. They are rules that describe and prescribe actions in particular circumstances. This is slightly obscured by the language of rights, because rights do not seem to imply statements in the standard "do X" or "if X then do Y" form. But rights and obligations are mutually interdependent and interdefinable. The claim that the Nuer have the right to use a piece of land implies that the Dinka must refrain from using it. So the assignment of rights does identify an equilibrium after all, in which the Nuer do not have an incentive to use the land that is south of the river and the Dinka do not have an incentive to use the land that is north of the river. Attaining this equilibrium and avoiding conflict is exactly the purpose (in the functionalist sense) of the introduction of property rights.

So overall the translation of constitutive rules into regulative rules does not cause any theoretical loss. There is room for functions and deontic powers in the unified theory. In a sense, it is obvious that this must be the case: how could a constitutive rule change people's behavior merely by introducing a new term? Searle admits that "the point of having deontic powers is to regulate relations between people" (1995: 100). To change behavior, then, the new term Y must indicate some actions that the players ought or ought not to perform. These rules must have the classic regulative form, "do X" or "if X then do Y," augmented by a deontic operator ("you *must* do X," or "if X then you *must* do Y"). So there must be some regulative rules hidden in the Y term, and with the help of a unified theory we can make them explicit. The rules-in-equilibrium approach shows in what way the deontic force of a rule can change the games that people play. Searle's insight that institutions make new types of behavior possible is vindicated in this account.

REFERENCES AND FURTHER READINGS

Most of this chapter is based on Guala and Hindriks (2015). Searle (2015) and Hindriks and Guala (2015b) offer a comparison between our perspectives on the normativity of institutions. On the problem of normativity in philosophy and social science, see Turner's (2010) excellent book, and on Searle in particular see Turner (1999). Miller (2010) proposes an account of institutions that emphasizes their normative foundations, from a standpoint that is very different from the one I take in this book. The twin ideas that the power of norms can be represented as perturbations of payoffs and that norms change the structure of the game are common in rational choice theory; see, for example, Ullmann-Margalit (1977), Sugden (1986), Pettit (1990), Bicchieri (2006). The article by Crawford and Ostrom (1995) offers a clear concise account, and Vanberg (2008) a critical view. I have argued that conventions have a tendency to turn into norms in Guala and Mittone (2010) and Guala (2013a). On the normative power of conventions see also Tummolini et al. (2013). Bicchieri (2006) is the most influential and comprehensive theory of social norms in the contemporary literature. Michael and Pacherie (2015) discuss normative commitments as uncertainty-reducing devices for social interaction. For different accounts of the source of normativity, see, for example, Sugden (1986, 1998a, 2000a) and Pettit (1990) on the desire not to frustrate others' expectations; Frank (1987), Gibbard (1990), and Nichols (2004) on emotions; Raz (1999), Skorupski (2010), and Broome (2013) on rational justification; Gilbert (1989) on joint intentions.

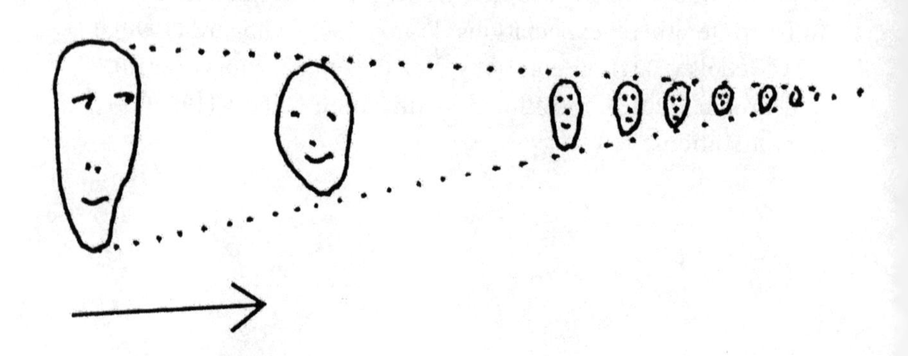

INTERLUDE

CHAPTER 7

MINDREADING

Humans are not the only creatures who are able to solve coordination problems. *Pararge aegeria*, the speckled wood butterfly, is genetically programmed to play bourgeois strategies in games of sexual competition, for example. But it can play just *those* strategies, and no other social arrangement is likely to emerge during the lifetime of an individual butterfly. Humans are more flexible and versatile. We can coordinate on different equilibria, and we can create new ones by inventing new correlation devices.

This is a tremendous advantage, but at the same time it creates difficulties that butterflies can ignore. A butterfly does not have to be assured that its competitors will do their part in the equilibrium that has been selected by evolution, because their behavioral repertoire is limited. There is no need to think about what the other players might do, so as to adjust one's behavior accordingly: nature takes care of coordination. For humans, in contrast, anticipating each other's moves is a vital necessity. Successful coordination requires not only that we all follow the same behavioral rules, but also that we believe that everyone is going to do their part, and that they believe the same. Since the attribution of mental states plays a central role in rule following, I will call it *the problem of mindreading* for coordination.

In games with multiple equilibria the optimal actions of the other players are not independent of our actions; what is best for

	L	R
L	1, 1	0, 0
R	0, 0	**1,1**

FIGURE 7.1. A coordination game with salience.

them depends on what we do, and what is best for us depends on what they do. A sophisticated agent should be aware of this mutual dependence. Jill cannot make a decision until she has identified what Jack has in mind, but Jack cannot decide until he has identified what Jill has in mind, and so forth in an endless circle. Sophisticated reasoning therefore may lead to trouble, preventing the formation of the beliefs that are necessary for coordination.

Nonetheless people do coordinate, frequently and reliably. So how do they do it? Schelling and Lewis argued that focal points help solve the problem, as we have seen. Consider the game in Figure 7.1, for example. Most people, just by looking at the matrix, notice that one of the outcomes is salient. It "stands out" in virtue of a logically irrelevant detail—the payoffs are printed in bold. Thus when they make their decisions, many players find it natural to choose R. They choose that strategy because they think that it will increase the probability of coordination. And they are right: by playing R, they improve significantly their chance to succeed.

This story can be generalized. In the grazing game, for example, the Dinka and the Nuer use the position of the old river as a correlation device, because that is the way in which the problem has been solved until now. The mechanism in this case is not essentially different from the one that helps solve the game of Figure 7.1, because the history of play is just another source of salience. The fact that a given profile of strategies has been chosen until now creates a focal point. And people decide (correctly) to converge on the focal point under the (correct) expectation that the others are going to do the same.

- I believe that X
- I believe that you believe that X
- I believe that you believe that I believe that X
- I believe that you believe that I believe that you believe that X

etc.

- You believe that X
- You believe that I believe that X
- You believe that I believe that you believe that X
- You believe that I believe that you believe that I believe that X

etc.

FIGURE 7.2. Common belief about X.

How can focal points solve the problem of mindreading though? For all the players to converge on the same solution, there must be mutual expectations of coordination. In theory, the players should form a complex system of interlocking beliefs: in a coordination game, if I expect you to choose X (to do your part in the equilibrium) then I must believe that you expect me to choose X. But then I must also believe that you believe that I expect you to choose X. And that you believe that I believe that you expect me to choose X, and so forth for every level of beliefs.

A case with two individuals is represented in Figure 7.2. When beliefs are nested in this manner, we shall say that there is *common belief* about X. Lewis—who introduced the concept—stated clearly that it is *not* meant to describe a reasoning process or an occurrent psychological state. The hierarchy of beliefs is not necessarily represented anywhere in anybody's mind. The nested structure provides only a conceptual analysis of the notion of common belief. Since coordination requires "nothing more nor less than intuitively perceived mutual expectations" (Schelling 1960: 71), the intuition may remain largely implicit.

Imagine for example that we are both sitting at the same table, and we see that there is a bottle of wine in front of us. I intuitively believe not only that there is a bottle and that you believe that there is a bottle, but also that you believe that I believe it, and so forth, without representing explicitly the whole belief structure. It is surely unrealistic to imagine that people entertain explicitly many levels of mutual expectations, at least in ordinary circumstances.

Where do common beliefs come from, however? How can a focal point create "intuitively perceived mutual expectations"? It is useful to have some general terminology to tackle this question. Suppose it is common belief in a population that a rule R (if X then do Y) is in place and that everyone will comply with it. Let us suppose further that the common belief is grounded in a *public event*, P, that *indicates* R (P→R). The public event is like the bottle being in front of us. In the case of the grazing convention in the Sobat Valley, the event is a behavioral regularity. The fact that the Dinka and the Nuer have behaved in such a way until now indicates a rule that projects the regularity in the future. Alternatively, the event may be a public statement, a ceremony, or a signal: somebody may say in public: "if X then do Y" which indicates R. In any case, the public event makes a profile of strategies salient.

P is called the *basis* for common belief in R, and the relation of indication must involve some kind of inductive inference. According to Lewis, if P indicates R, then there is common belief about R. The publicity of P and the relation of indication generate common belief about R.

Although Lewis's theory accounts for many ordinary cases, however, it is not general enough. The problem is that some institutions, the conventions of traffic for example, are perfectly robust even though they do not seem to be grounded on any observable event. The theory therefore seems to require too much. Ken Binmore, who has been a vocal critic of common belief assumptions, puts it like this:

> How often do we have the opportunity to observe each other observing something? For large numbers of people, I guess the answer is never. So how can a language have become a convention if a convention needs to be common

> knowledge in a society? How can it have become conventional for gold to be valuable? How can it even have become conventional to drive on the right? And if we do not know the answer to such questions as these, how are we ever to find our way to an equilibrium of the game of life we play on this planet in which we get global warming and the like under control? This is not to deny that we behave as if conventions are common knowledge when we use them, but Lewis's definition would seem to make it impossible for conventions to get started except in very restricted circumstances. (Binmore 2008: 23)

Evolutionary game theorists like Binmore prefer to make do without beliefs altogether. They advocate the use of models with belief-less players, endowed with minimal cognitive capacities. But even those who prefer to work with cognitively rich models must admit that the point is well taken. Indication cannot ground coordination with strangers, if no public event constitutes the basis for the formation of common beliefs.

Another worry concerns the notion of indication itself. Lewis is careful to say that for indication to work the agents must apply the same "rationality, inductive standards, and background information" (1969: 53). R can be grounded in P only if the players are *symmetric reasoners*: they must make the same inference from P to R, and be assured that the others do the same. But why should they believe that they all reason symmetrically? Presumably there must be some sort of grounding for the assurance that P indicates R. And what could that be? Presumably we need another public event P' that indicates that P indicates R (P'→(P→R)). But what could *that* event be?

So we have two problems to solve: first, we need a story about the creation of common beliefs that does not necessarily involve the observation of a public event; second, we must explain what the relation of indication rests upon. I will refer to these two issues jointly as the *grounding problem* for common beliefs.

A clue to solve the second part of the problem can be found already in Lewis's text. Lewis at one point gives an account of belief attribution that does not involve any regress, but relies instead on a process of empathic understanding. We may try to find out what the other players believe by trying to think in the same way as they do.

> We may achieve coordination by acting on our concordant expectations about each other's actions. And we may acquire those expectations, or correct or corroborate whatever expectations we already have, *by putting ourselves in the other fellow's shoes*, to the best of our ability. If I know what you believe about the matters of fact that determine the likely effects of your alternative actions, and if I know your preferences among possible outcomes and I know that you possess a modicum of practical rationality, then I can *replicate* your practical reasoning to figure out what you will probably do, so that I can act appropriately. (1969: 27)

The italics in this paragraph are mine, to emphasize two important notions: putting oneself in someone else's shoes, and replication. Putting oneself in the shoes of another person is to adopt her own perspective, for example, by endorsing her goals and representing the world as she sees it. To replicate someone's reasoning is to follow the same procedure she is using to reach a conclusion or to make a decision. It involves the *simulation* of somebody else's thoughts.

I take this paragraph to suggest that, according to Lewis, indication is not a theoretical inference. The mutual expectations that are necessary for coordination are not derived from a set of theoretical assumptions concerning the effect that some event P will have on other people's beliefs, but by going through the same procedure they use to form their beliefs, that is, by symmetric reasoning. The distinction between theoretical inference and simulation is largely implicit in Lewis's text, which is not surprising, because it would emerge clearly in the philosophical literature only two decades later. Today we refer to these two approaches as the "theory-theory" and the "simulation theory" of mindreading.

Some versions of game theory adopt tacitly a theory-theory of mindreading: each individual must form beliefs about the behavior of the other players, and in so doing she is supposed to use the same procedures that game theorists use to explain *her* behavior. Each player figures out what the preferences and beliefs of the other agents are, uses rational choice theory to predict their behavior, and then chooses an action that maximizes her own payoffs, given their expected behavior. The underlying presumption is that people predict others' behavior in the same way as a scientist predicts the movement of a physical object: start from a theory with laws and free variables ($f = ma$), fix some parameters (mass, force), and derive a prediction (acceleration). In the case of human behavior, the free variables are preferences and beliefs, and the theory is typically a theory of rationality that relates preferences, beliefs, and choices in a systematic way. For example, if I know that Jill prefers A to B, and that she believes that doing X will lead to A and not doing it will lead to B, then I predict that she will do X.

The attribution of preferences is usually considered nonproblematic in coordination games. We all know what the possible equilibria are. But in order to attribute beliefs, from a theory-theory perspective, we must try to figure out what information the other players have about the situation. Then, assuming normal reasoning capacities, we can try to infer their behavior. The problem is that in a coordination game the actions of each agent depend on her beliefs about the actions of the other agents, and vice versa. So we cannot fix one variable until we have fixed the others.

Inferring beliefs from the observation of past behavior will not do either: the other players can legitimately expect that we will continue to do what we did in the past only if they think that we expect them to do the same. But how can they believe that, if we do not know what to think? Any conjecture based on observation is going to be unjustified. Another way to put it is that game theory, if it is based on a theory-theory of mindreading, cannot issue a precise prediction because its model is underdetermined by the available data.

But people do solve coordination games, so they must predict each other's behavior in some other way. According to the *simulation theory* of mindreading, when we attribute mental states to other agents we do not apply a theoretical framework with preferences and beliefs as main variables. Mindreading is less a matter of theorizing, and more a matter of doing—of replicating, in particular, the reasoning of other agents.

How can simulation solve the problem of indication, and explain the formation of common beliefs? Lewis says that two inputs are needed to replicate your reasoning: I must know your preferences, and I must know "what you believe about the matters of fact that determine the likely effects of your alternative actions." The first assumption is unproblematic, if we know that we are playing a coordination game. But the second assumption is not. Among the matters of fact that determine the effects of your actions there are, crucially, my actions. So Lewis is asking that I know what you believe about my actions, but how can I know it? Belief attribution was the problem that we were originally trying to solve!

Adam Morton, one of the pioneers of simulation theory, has proposed a simple account of mindreading in coordination games. The idea is that when two individuals have a goal in common they predict each other's actions and interpret each other's beliefs by means of a procedure that he calls *solution thinking*. Each agent asks: what is the easiest or the most natural way to tackle this problem? What is the obvious solution? If there is a clear answer, then the same reasoning is attributed to the other player by default.

> One first thinks of an outcome which one can imagine the other person or persons both would want to achieve and

> would believe that one would try to achieve. One then thinks out a sequence of actions by all concerned that will lead to it. Lastly, one performs the actions that fall to one's account from this sequence . . . and expects the other(s) to do their corresponding actions. (2003: 120)

Let us examine the various steps in detail. Seen from the point of view of an individual player, solution thinking can be reconstructed as follows:

1. S is the obvious solution to the coordination problem.
2. You also think that S is the obvious solution to the coordination problem.
3. To achieve S, I must do X and you must do Y.
4. You also think that I must do X and you must do Y.

Simulation occurs in the second and fourth steps. At step 1, I look at the problem and identify a focal point (the "obvious solution"). Step 2 replicates the procedure for the other player: she identifies the same focal point because she is just like me. Once the solution has been identified, I can derive my own actions and the actions of the other player by simple instrumental reasoning (step 3). Using the same procedure ("she reasons in the same way") finally I predict what she will do and what she believes that I will do.

> [Solution thinking] is simulation. First of all it results in an understanding of others: you have an idea what the other is going to do, and you have the materials for putting together an account of why they might do it. And the understanding is got by reproducing rather than representing their thinking. (Morton 2003: 120)

Solution thinking satisfies the principle of symmetric reasoning postulated by Lewis, but the grounding regress does not arise: your beliefs about my actions are derived by simulation. This solves the problem of grounding the indication relation. We do not have to justify theoretically the inference from P to R, because indication is not a theoretical inference in the first place. It is a simulation procedure, which does not have to be further justified. The question is whether it works, not whether it is based on firm theoretical assumptions.

But simulation solves the other half of the grounding problem as well, because simulation does not necessarily move from a public event. Consider one of the classic examples of salience discussed by Schelling in *The Strategy of Conflict*. In his informal experiments, Schelling used to present the following scenario:

> You are to meet somebody in New York City. You have not been instructed where to meet; you have no prior understanding with the person on where to meet; and you cannot communicate with each other. You are simply told that you will have to guess where to meet and that he is being told the same thing and that you will just have to try to make your guesses coincide. (Schelling 1960: 55)

Where would you go, and when? The most common answer was the Grand Central Station at noon. Clearly the answer was not reached by trying to figure out what the other person could possibly believe that one believed, and so forth. Such a procedure would lead to the usual regress. Schelling's subjects instead attributed to the other individual the belief that they would go to Grand Central in the same way as *they* formed the belief that Grand Central is the place to go. Among the infinitely many solutions to this problem, Grand Central struck them as the most natural and obvious solution—the one that first came to mind—and the common belief was derived from that insight alone.

Notice that there is no public event P to ground the common belief. The basis in fact is the *opposite* of a public event: it is the very private feeling that Grand Central is the obvious solution. I believe that you believe (etc.) that X, because X is obvious to me, and you are just like me.

Morton uses solution thinking to make a general philosophical point about mindreading. He argues that the attribution of beliefs is often an effect of our desire to coordinate. In many circumstances we do not coordinate because we have the same beliefs, but rather we have the same beliefs (we attribute one another the same mental representations) because we want to coordinate. The trick is to infer others' beliefs from a common target, assuming tacitly that the other players are just like us. Beliefs then will converge: everyone will end up attributing to others the same beliefs that she attributes to herself.

One of course can ask whether using this procedure is appropriate, whether it is likely to produce reliable results. And the answer is: not always, but *in normal conditions* simulating is a remarkably efficacious way of solving coordination problems. We simulate others, they simulate us, and it usually works.

What does "normal" mean? It means that the agents (the simulators) must be attuned to each other's thinking. They must share much in terms of biological dispositions, cultural background, factual information, education, and reasoning capacities. But people do not need to believe that this is the case *before* they start simulating—they do not have to endorse a theory that implies this claim as one of its consequences. Usually, they take symmetric reasoning to be the default condition. Only when people have some reason to believe that the symmetry may *not* hold, their disposition to simulate is inhibited. They stop simulating and start theorizing, asking sophisticated but awkward questions about others' beliefs and desires.

Imagine you are on the road, in France say, and you see another car swerving from right to left. You look at her plates and notice that the car is registered in Britain. This case is abnormal and simulation is put on hold. You are aware that what you ought to do depends on what the other driver will do, and that the latter depends on what she thinks that you will do, but you have no way to figure that out ("Does she know that she is in the wrong lane? Or does she think that I am in the wrong lane?"). In these circumstances clearly coordination is at risk, because you have to think as a game theorist. The fast track to mindreading (simulation) is blocked, and you engage in unnatural habits of thought. Notice that this is not only unusual—it is very difficult. When we teach game theory, we force students to engage in an unusual way of thinking, which they find hard, and for good reasons. Thinking about higher order beliefs often hampers rather than facilitates coordination.

So the claim is not that simulation is the only true account of mindreading. There are many styles of mindreading and we do engage in theoretical reasoning about others' beliefs in some occasions. But in a number of circumstances—that are important and common—we adopt a different style of prediction based on simulation. And this style helps solve problems of coordination.

REFERENCES AND FURTHER READINGS

Lewis (1969) and Aumann (1976) are usually credited for having introduced the concept as well as the expression "common knowledge" in the literature; Lewis however later regretted using that term, and acknowledged that "common belief" is preferable. An overview of the literature can be found in Vanderschraaf and Sillari (2013). The formulation of common belief that I have used in the text is not entirely faithful to Lewis, who preferred a formulation in terms of "reasons to believe" (see Cubitt and Sugden 2003). The notions of "basis" and "indication" are in Lewis (1969). On common beliefs and conventions see also Sillari (2008) and, for a skeptical perspective, Binmore (2008). Evolutionary models of coordination and cooperation are illustrated and defended by Skyrms (1996, 2004) and Binmore (1994, 1998, 2005). On the importance of symmetric reasoning and shared inductive propensities, see also Sugden (1998b). There are many versions of simulation theory; Goldman (1989, 2006) and Gordon (1986) provide classic statements. The discovery of "mirror neurons" has made the theory increasing popular among scientists and the general public (see Rizzolatti and Sinigaglia 2008). The idea that simulation offers a solution to game-theoretic problems of coordination has been put forward by Morton (1994); I recommend his unusual and suggestive discussion in chapters 3 and 5 of *The Importance of Being Understood* (2003).

CHAPTER 8
COLLECTIVITY

Solution thinking is an individualistic mode of reasoning. Mental representations are attributed to individual players, and are expressed in the classic "I believe," "you believe," "she believes" format. In this respect, Morton's account differs from those theories in social ontology that emphasize the role of mental states expressed in collective mode ("we believe," "we prefer," and so forth). Such theories have become increasingly influential in recent years, and are generally known as theories of *collective intentionality* (other expressions, like "shared" and "joint" intentionality, are also commonly used in the literature, but I will treat them as synonymous).

Intentionality is a technical term used in philosophy to refer to the property of a mental state to be about something or to be directed toward something. For example, if I believe that John's trousers are too big, then my mental state is about the size of his trousers; if I want to have an espresso, then my mental state (a desire, in this case) is about drinking a cup of coffee. Among the mental states that have intentionality, there are also *intentions* in the ordinary sense of the term, that is, mental states that play a specific role in directing and causing action (for this reason, and to avoid confusion with the more general concept of intentionality, philosophers sometimes call them "intentions in action"). An intention in action is a purpose to do something—

roughly, a plan that guides one's behavior, like for example my resolution to finish this chapter before the end of the week.

Intentionality is a notoriously difficult and controversial topic, but luckily for our purposes a rough definition is enough. The debate on collective intentionality starts from the observation that intentional states can be expressed in individualistic as well as in collective format. For example, it is perfectly legitimate in grammatical terms to say that *we* want to go on holiday in Greece, where the desire is attributed to a collective constituted by my wife, my children, and I; or that *we* believe that the recent cuts in the higher education budget will affect the quality of teaching, where the belief is attributed to the lecturers of my university.

To posit collective intentional states raises several interesting puzzles, though, which over the years have enticed an increasing number of philosophers and cognitive scientists. Who or what is the carrier of a collective mental state? Is it rational for an individual to intend an action that she cannot possibly carry out by herself? Do collective intentional states constitute a separate type of cognitive states, or are they reducible to individual mental states?

Notice that most of these puzzles seem to be related only indirectly with the philosophy of social science. Standard social science does not posit the existence of collective intentions, and this seems to suggest that a large number of phenomena can be explained without it. Nonetheless, the problem of collective intentionality has become a central and preliminary chapter in many discussions of social ontology. Many philosophers think that the current neglect of collective intentions imposes limitations on social scientific explanations, and that philosophers can contribute to redress this state of affairs. As a consequence they have argued—or have implicitly assumed—that the best approach to understand institutions is to start with an analysis of collective intentionality, and then build a social ontology upon it.

These philosophers typically claim that collective intentions are a *necessary* element of sociality. According to Raimo Tuomela (2002a: 183), for example, "we-mode collective acceptance creates, and is required for, institutional entities and practices": "Central social notions as cooperation, social institutions, and the evolution of institutions, as well as collective and group responsibility, do require the full we-perspective as an underlying notion" (2002a: viii). John Searle similarly uses collective intentionality as a criterion to distinguish social from nonsocial facts: "Any fact involving collective intentionality is a social fact. Thus, for example, hyenas hunting a lion and Congress passing legislation are both cases of social facts. Institutional facts . . . are a special subclass of social facts" (1995: 38). What distinguishes social from nonsocial phenomena, according to these philosophers, is a feature of the agents' mental states. This in turn implies that progress in social ontology depends on the solution of various difficult problems in the philosophy of mind—problems that are still hotly debated and that so far seem to be impervious to empirical investigation.

One wonders whether this is a reasonable approach. Intuitively, it seems wise to give first an account of sociality and social institutions that relies as little as possible on controversial notions such as collective intentionality. Then, once the main features of our ontology have been outlined, we may proceed to ask what kind of psychology is necessary for institutions to exist. Should it turn out that collective mental states are not necessary, this "transcendental" procedure would not demonstrate conclusively that they do not exist, of course. (In fact, as we shall see, there are good reasons to believe that they are an important enabling condition for sociality.) But this strategy will help us avoid making our theory dependent on their existence—hostage to a debate that takes place in another, equally if not more controversial area of philosophy.

So, why should one believe that collective intentionality is a necessary feature of sociality? The classic motivation is that there are features of social or institutional phenomena that cannot be explained if we attribute individualistic intentions to the participating agents. Searle for instance has devised a famous thought experiment—the "Business School" example—to show that no set of individual intentions, no matter how interlocked, is able to account for certain forms of coordinated action. The thought experiment begins with the story of a group of graduates who have been indoctrinated to believe in Adam Smith's invisible hand theory, that is, the hypothesis that one can benefit humanity simply by pursuing one's own selfish interests. So each business school graduate decides to pursue the same goal (helping humanity) individually, without cooperating with anybody, but knowing that every other graduate will do the same, and that they all know it.

This, according to Searle, is a paradigmatic social phenomenon that can be explained by appealing to a set of standard beliefs and desires expressed in I-mode. If the story is modified only slightly, however, the same explanation becomes inadequate:

> This case has to be distinguished from the case where the business school graduates all get together on graduation day and form a pact to the effect that they will all go out together and help humanity by way of each pursuing his own selfish interests. The latter case is a case of collective intentionality the former case is not. (1990: 405)

What is the difference between the two stories, exactly? Searle argues that

> the notion of a we-intention, of collective intentionality, implies the notion of *cooperation*. But the mere presence of I-intentions to achieve a goal which happens to be believed to be the same goal as that of other members of a group does not entail the presence of an intention to cooperate to achieve that goal. (1990: 406)

The notion of cooperation is, unfortunately, not much clearer than the notion of collective action, but here and elsewhere Searle draws a strong link between cooperation and the notion of a "pact"—an agreement or commitment to achieve a common objective. This notion is central also in other versions of collective intentionality theory. According to Margaret Gilbert (1989), for example, a group of people who engage in a collective action are joined by a commitment (a "quasi-agreement," in her terminology) to achieve a goal. And this joint commitment is a necessary feature of any structured form of sociality—from customs to traditions, conventions, and even the laws of a given community.

> Our conception of a social convention is the conception of a quasi-agreement which is generally regarded as grounding an ought judgment, where it is common knowledge in the population that this is so. (1989: 369)

A large portion of Gilbert's seminal book, *On Social Facts*, is devoted to criticize Lewis's theory of conventions. Gilbert complains that an individualistic theory like Lewis's is unable to account for the main features of social conventions as they are ordinarily conceived. And the feature that is given most prominence in Gilbert's text is their intrinsic normativity—the fact that the members of the relevant group are bound to follow the convention by a joint commitment. Gilbert goes as far as to say that "the link is a conceptual one: normative attitudes are to social conventions as femaleness is to sisterhood" (1989: 350).

The focus on normativity is a common characteristic of collective intentionality theories. The essay by Wilfrid Sellars that is generally regarded as the forerunner of the contemporary literature is essentially about morality; and the same concern with normativity can be found in the work of Tuomela and others. The common idea is that individuals are committed to certain behaviors in virtue of the fact that they have a we-mode intention. Collective intentionality is needed because ordinary

I-mode intentions, no matter how interlocked, do not suffice to generate such a commitment.

This position, as I said, is controversial. Some philosophers (like, notably, Michael Bratman) have offered accounts of collective actions that do not emphasize normativity and that do not drive a wedge between individual and collective intentional states. But, be that as it may, we do not need to take a position on this matter. Let us suppose, for the sake of the argument, that the normative element of social institutions cannot be accounted for by positing individualistic mental states. Even if this were true, the thesis of the necessity of collective intentions would fail for another, deeper reason: the distinction between collective and individual mental states does not mirror the distinction between institutional and noninstitutional, or even social and nonsocial phenomena. A taxonomy based on collective intentions does not cut the social world where it should.

To see why, it is sufficient to point out that many social institutions do not rely on strong normative commitments such as those that are supposedly engendered by a joint intention. Consider for example racist institutions such as those that used to be common in many European, American, and African countries until not so long ago. Such institutions were constituted by many rules, some of which were encoded formally in the legal system while others were enforced by other means. In South Africa, for instance, black people were expected to use different public transportation and to live in segregated townships, and white people could decide which land the various "races" or ethnic groups could use. Informal conventions were also an important part of the apartheid system. In his autobiography, *Long Walk to Freedom*, Nelson Mandela recalls that at tea time the black lawyers of the firm where he worked were expected not to use the same cups used by the white employees.

Now, to say that these rules were based on a collective agreement seems rather strange. The rules certainly carried deontic force: the black person who refused to conform was breaching a norm. Those who conformed in turn did it because they were expected to do it and did not dare contradict these expectations, because they feared the punishment inflicted on deviants. But the norm did not rest on a collective agreement or a group intention: the whole point of apartheid in fact was to *undermine* the notion that black and white people belonged to the same community.

The underlying problem is that normativity is an ambiguous notion. On one reading, a normative behavior is just a behavior that is expected and sanctioned in the community. This is the standard social science interpretation, and the one that I have used in chapter 6. On another interpretation, a normative behavior is mandatory in virtue of a stronger binding relation—like a pact or a promise—that has been forged in the community. The normative force, in this case, is based not merely on expectations but on some underlying contract that provides legitimacy to the sanctions.

Returning to our example, it seems clear that the racist institutions of South Africa were endowed with normative power in the former, but not in the latter sense. And yet surely they were genuine institutions. But then we are forced to admit that normativity in the sense of joint commitment is not a necessary feature of institutions. Whether collective intentions are involved or not in social action is a contingent issue, and we should better not build social ontology on this doubtful premise.

Some collective intentionality theorists have recognized the problem, and over the years have modified their position accordingly. For example, Searle now claims that institutions require only collective "recognition," instead of collective acceptance of a rule. This means that some institutions, like apartheid, exist even if some people are not bound by a commitment and do not consider the rules legitimate: "one can recognize and act within institutions even in cases where one thinks that the institution is a bad thing" (Searle 2010: 57). But with commitment, collective intentionality also has to go.

> This is an important point, because it shows that there are some forms of collective intentionality which are reducible to I-intentionality plus mutual belief. If you have collective recognition of something as money, that collective recognition can be constituted by the fact that each person recognizes money and there is mutual knowledge among participants that they recognize money. (2010: 58)

So this line of argument, if correct, demonstrates that collective intentionality is not necessary for the existence of institutions in general. It does not demonstrate that it does not exist, however. As I pointed out earlier, there is a big difference between proving that something is unnecessary and proving that it is useless. In fact, I think there are good reasons to believe that reasoning in collective mode is useful for coordination, and hence for the existence and resilience of institutions.

To see why, we have to go back to simulation and solution thinking. Simulation, intuitively, solves the problem of coordination by reducing the gap between the mental states of the simulator and the mental states that are attributed to the simulated individual. The problem of coordination is solved by "merging," so to speak, the two minds or using one to figure out what is going on in the other. Put this way, there are striking similarities between simulation and collective reasoning. In fact, I suspect that any creature who is able to engage in solution thinking is also able to think in a collectivistic way, because the distance between these two modes of reasoning is relatively small.

In particular, there is a strong family resemblance between Morton's solution thinking and *team reasoning*, a theory of coordination that has been developed by Michael Bacharach, Robert Sugden, and Natalie Gold. According to Gold and Sugden, the theory

> seeks to extend standard game theory, where each individual asks separately "What should I do?" to allow teams

	H	L
H	2, 2	0, 0
L	0, 0	1, 1

Solution thinking:

- I want to earn 2 dollars;
- She wants to earn 2 dollars;
- The only way to earn 2 dollars is by choosing H;
- I will choose H;
- She will choose H.

	H	L
H	4	0
L	0	2

Team reasoning:

- We want to earn 4 dollars;
- The only way to earn 4 dollars is by choosing H;
- We will choose H.

FIGURE 8.1. Simulation (solution thinking) versus team reasoning.

of individuals to count as agents and for players to ask the question "What should we do?" This leads to team reasoning, a distinctive mode of reasoning that is used by members of teams, and which may result in cooperative actions. (2007a: 110–11)

So an important difference between team reasoning and collective intentionality theories is that the former, but not the latter, focuses on the process that leads to the formation of an intention to act. The question of what makes collective intentions distinctive as mental states, therefore, is temporarily set aside to focus on how mental states formulated in we-mode may facilitate the solution of problems of coordination that cannot be satisfactorily addressed within an individualistic framework.

Take the puzzle of hi-lo for example (Figure 8.1, on the left). Hi-lo presents all the usual problems of coordination games, including the problem of forming mutual expectations. Even though we both prefer to coordinate on HH, I should choose

High only if I believe that you will choose High; but you should choose High only if you believe that I will choose High, and neither of us can be assured that the other will do it because our decisions and beliefs are interdependent. Nevertheless, most people find hi-lo utterly trivial: the right thing to do, clearly, is to choose High.

Team reasoning theorists argue that the game looks trivial because it is perceived as a parametric problem for a single agent, the team constituted by the two players, rather than as a problem of strategic interaction. The first step to solve the problem, then, is to look at it from the point of view of a single agent, the team. To make the point in the simplest manner, let us suppose that the payoffs of the team are obtained by simply adding the payoffs of the two players. From the perspective of the team, hi-lo then looks like the transformed matrix on the right of Figure 8.1.

Below each matrix I have added a possible line of reasoning leading to the "obvious" solution. The one on the left is the mode of reasoning based on solution thinking. The other one is team reasoning. In solution thinking each player replicates the other player's reasoning using oneself as a model and then proceeds as if the problem of belief interdependence did not exist. The solution that appears obvious (HH) is automatically used to identify a strategy, which is then used to predict the behavior of the other player. The premise in this case is a standard individualistic goal ("I want to achieve HH").

In team reasoning in contrast the point of departure is a premise formulated in we-mode ("we want HH"). Since the payoffs of the two players are perfectly aligned in hi-lo, team reasoning delivers the same result as solution thinking. But this is not a matter of coincidence, for these two reasoning strategies have very similar properties. In both cases a problem of coordination is solved by taking the perspective of another agent. In both cases the solution is identified using oneself to simulate someone else's reasoning. In both cases the reasoning proceeds

	White first	Black first
White first	6, 3	0, 0
Black first	0, 0	3, 6

FIGURE 8.2. A coordination problem with unequal equilibria.

backward from a solution to the strategies that help attain that solution, and in both cases the game is transformed into a parametric decision.

There are some differences between team reasoning and solution thinking, though. The first difference is that in team reasoning HH is the obvious solution because it is the best one for the team. The rational solution is the focal point. In solution thinking, in contrast, HH is focal because it is the best outcome for me and the best outcome for you. Since we both think in the same way, HH is the obvious thing to do.

An advantage of solution thinking, then, is that it is applicable to all focal points. Take for example a coordination problem with two very asymmetric equilibria, such as the one in Figure 8.2. The idea that the players think as a team seems a priori rather implausible, and empirical data indicate that payoff inequalities tend to undermine coordination in games like these. However, suppose that one of the two equilibria has been made salient by history, for example because one of the players is identified by a biological marker (sex or race) that has been used for centuries as an indicator of morality, power, and privilege. In this case, it would not be surprising if the players coordinated smoothly on the focal point solution.

This example is not far-fetched–it is in fact what happens regularly with sexual and racial discrimination. And as we have seen in the last chapter, solution thinking is able to account for this kind of coordination. Solution thinking is more general than team reasoning, because it is applicable to all focal point effects, and not just to situations (like hi-lo) where one outcome is socially optimal and there is no conflict of interest between the players.

Notice that solution thinking and team reasoning also provide different explanations of coordination failures: on the one hand, coordination may fail because I cannot be confident that the others see the same obvious solution as I do (it is a simulation failure); on the other, it may fail because there is no "fusion of egos" in the collective agent. But the two processes may well be complementary, and be used by the same individuals depending on what the situation requires. In the case of team reasoning the agent is a third party—the team—with which the players identify, and the reasoning of the team is simulated in we-mode. In solution thinking, in contrast, all the thinking is carried out in I-mode; the players identify with one another, and their beliefs are derived by reasoning backward from an outcome to a profile of strategies. The principle of symmetric reasoning is used to generate some key premises, that is, to identify the "right" equilibrium solution and to assure both players that they expect each other to do their part in the equilibrium. But there is no big difference between using me to simulate you, or using me to simulate an entity that includes me and you. In the end, I suspect that the style of reasoning is what really matters for coordination; whether we start from an "I" or a "we" may well be a secondary detail.

REFERENCES AND FURTHER READINGS

The philosophy of collective intentionality stems from the work of Sellars (1963). The classic statements of the theory are by Tuomela and Miller (1988), Gilbert (1989), Searle (1990), Bratman (1993, 2014), Tuomela (1995, 2002a, 2007). Team reasoning has been proposed and developed by Sugden (1993, 2000b, 2003), Bacharach (1999, 2006), Gold (2012), Gold and Sugden (2007a).

For a comparison between collective intentionality and team reasoning, see Bardsley (2007), Gold and Sugden (2007b), Hakli, Miller, and Tuomela (2011). In psychology, there is an older body of experimental work on so-called group identity (Tajfel 1982; Tajfel and Turner 1986) as well as a new stream of empirical research on joint action (reviewed by Gallotti and Frith 2013). Bianchin (2015) discusses the relationship between joint action and simulation, Turner (2010) the one between collective intentionality and normativity. Vromen (2003) draws on evolutionary biology to argue that collective intentionality is not needed to explain cooperation and coordination.

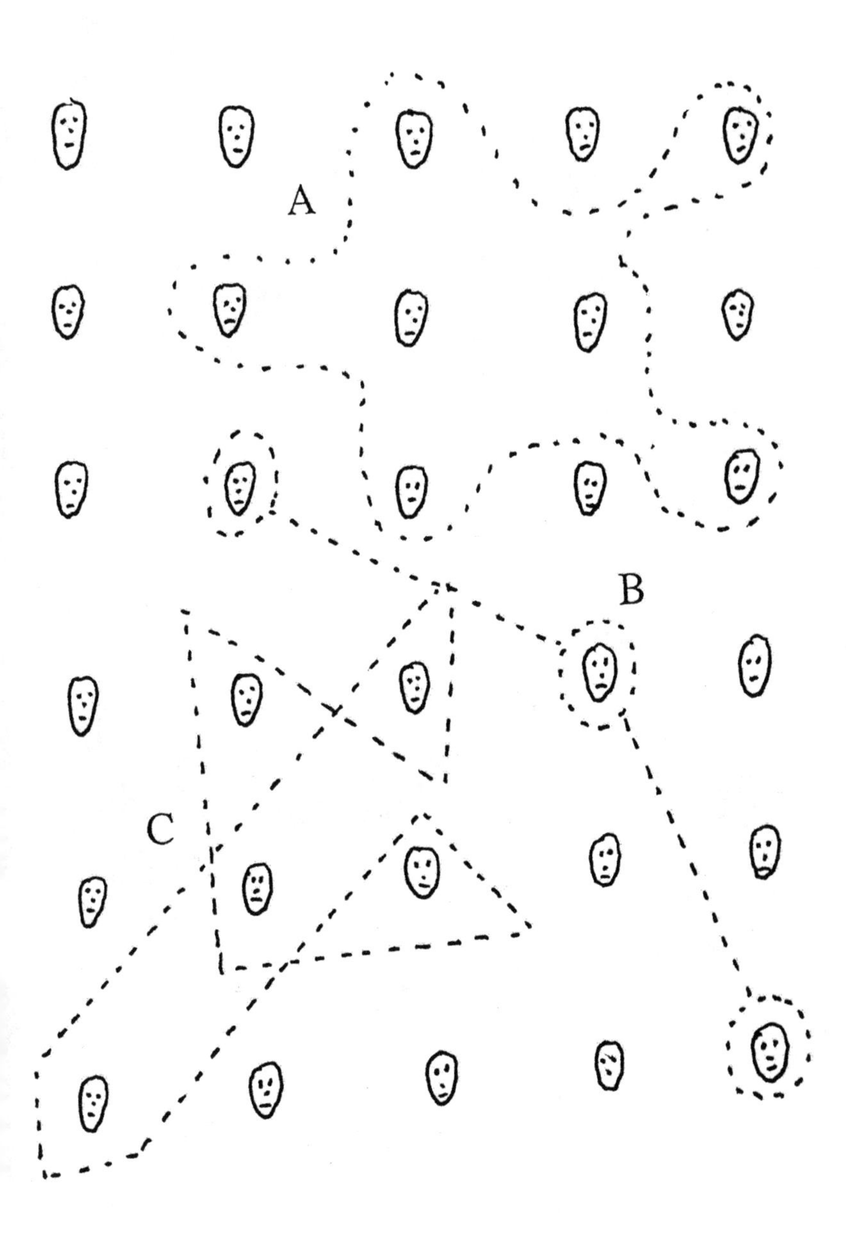
A
B
C

II

APPLICATION

CHAPTER 9
REFLEXIVITY

Institutions change the way in which we see the world. They do it in many ways: first of all, they change our beliefs about other people's behavior. Because the British rules of traffic prescribe driving on the left, we expect drivers to keep left. Because the garden in the back of my house is private property, I do not expect my neighbors to use it for their barbecue parties. Such beliefs moreover are tinged with normative colors: I would be surprised if my neighbors occupied my garden without permission, I would find it unacceptable, and I would probably try to redress the injustice.

But institutions change our worldview also in a less direct way: the theoretical terms that we introduce by means of constitutive rules create new ways of classifying people and things. Institutional terms are used to label people, for example when we assign roles that help coordination in our everyday interactions (husband, wife; policeman, thief; prime minister, king). And they are also employed to label things, when such things are used as correlation devices in the games that we play. A stone becomes a second base, a square line a penalty box; a piece of paper becomes a dollar bill, a building a church.

The two mechanisms are related, as we have seen in the first half of the book. The terms that we use to name institutional entities ultimately refer to sets of rules, and as such they are also tinged with normative color. There are things that you can do as a policeman but not as a judge; permissible behavior in your house that would be quite inappropriate in a church; and things that you can say to your children but that would be out of order when you speak to the king. Since the way in which things are classified influences the way in which people behave, naming seems to have the peculiar power of making things happen—of creating a reality that did not exist before.

This power appears less mysterious if we consider the mechanics of coordination. Since institutional terms refer to sets of regulative rules, naming itself must be understood as a strategic tool. Names ultimately refer to behaviors, and at the same time may change the behaviors that they describe.

The most common term used in scholarly conversation for this phenomenon is *reflexivity.* Its core meaning comes from mathematics: a reflexive relation ("is equal to," for example) relates every element of a set to itself. In social theory however reflexivity has been mainly used in the context of causal, rather than logical or mathematical relations. In a reflexive social relation, each event or variable is both a cause and an effect of another event or variable (X causes Y and Y causes X). There is also an affinity with the phrasal verb "to reflect on," meaning "to think carefully" on a given topic. So when sociologists like Anthony Giddens speak of "reflexive modernity," they refer to our unprecedented level of awareness about social organization, and the way it influences modern societies. Social phenomena are reflexive in part because people are reflexive in this sense.

The debate on reflexivity predates the introduction of this term. The single most influential contribution is Robert K. Merton's essay "The Self-Fulfilling Prophecy" (1948), where the expression "reflexivity" does not even appear. Merton's essay is rich of insights and suggestions that were developed separately by different scholars. One suggestion concerned the demarcation of natural from social science: Merton claimed that self-fulfilling prophecies mark an important divide between the natural and the social sphere (they are "not found among predictions about the world of nature"). Philosophers immediately challenged this claim: Adolf Grünbaum (1956) pointed out that analogous phenomena can be found in the physical and biological realm. His main example was the feedback mechanism of a missile-control system. Suppose that a computer is monitoring the trajectory of a rocket flying toward a moving target, and continuously sends predictions (feedback) to the rocket itself. Whenever the rocket seems bound to miss the target, the control system will send a warning, and the rocket will adjust its trajectory. But as long as the computer predicts that the target will be hit, the rocket will stick to its current path. It will hit the target (in part) because it has been told that it will hit the target.

Grünbaum's example was persuasive, and became a classic in the neopositivist literature of the 1960s. There is a discussion of reflexivity in Ernest Nagel's influential textbook, *The Structure of Science* (1961), that goes roughly along the same lines. And in the third edition of *Social Theory and Social Structure* (1968) Merton endorsed the neopositivist position. So by the 1970s reflexivity seemed to be fully assimilated in the philosophy of science.

But in the 1990s the reflexivity debate was revived again, under a different name. Ian Hacking introduced a new terminology, speaking of the "looping effect of human kinds." Looping effects are caused by feedback mechanisms, Grünbaum-style. But Hacking endorsed a form of demarcationism, Merton-style: feedback loops distinguish the human from the natural sciences.

> Responses of people to attempts to be understood or altered are different from the responses of things. This trite fact is at the core of one difference between the natural and human sciences, and it works at the level of kinds. There is a looping or feedback effect involving the introduction of classifications of people. New sorting and theorizing induces changes in self-conception and in behaviour of the people classified. These changes demand revisions of

> the classification and theories, the causal connections, and the expectations. Kinds are modified, revised classifications are formed, and the classified change again, loop upon loop. (Hacking 1995: 370)

Hacking's terminology reflects a shift in the debate. Classification, rather than forecasting, is the main focus of discussion. Ontology, rather than methodology, comes to the fore. Hacking tries to defend a position called "dynamic nominalism," the idea that people and objects in the social world do not come sorted into fixed kinds. The entities that social scientists classify change continuously, and in part they do because of the way in which they are classified.

The demarcation built on feedback loops reflects a significant divide for Western culture, the boundary between the world of nature and the world of men. It suggests that what can be known about humanity, and the way in which we can know it, differs considerably from the knowledge that we can gather about the natural world. The deep conviction that a demarcation exists goes a long way to explain the popularity and resilience of reflexivity in philosophical debates. Because the difference is so obvious, because it *must* be there, the issue of reflexivity keeps coming back in different guises and under different names. So it is important to examine carefully what kind of implications classificatory loops may have. The issue is sufficiently intricate to keep us busy for a few chapters: it involves the possibility of making predictions in the social sciences, the independence of social phenomena from social theories, the fallibility of folk and scientific theories of institutions, and the reality of social kinds. We will have to proceed gradually, addressing the simplest questions first. In this chapter I will just elucidate the notion of feedback loop, or how institutional terms may describe and change behavior at the same time. Social kinds will be the main topic of chapter 10, while realism and fallibilism will be discussed in chapters 11 and 12.

The mechanics of reflexivity is fairly simple: we represent X as Y, and as a consequence X begins to behave like an entity of type Y. Classifying X as Y contributes to her becoming Y. There are some famous examples of this kind of loop in the psychological literature: in the so-called Pygmalion effect; for example, labeling students influences their academic performance. In particular, the children labeled as good students perform better than those labeled as average or bad students, irrespective of their actual capacities. The causal loop in this case goes from classification to performance, and takes the form of a self-fulfilling prophecy.

The details of the mechanism may vary from case to case, depending on the intermediate causal processes that link classification to performance. For example, (1) the teachers may invest more energies in teaching "good" students, (2) they may be biased when they assess them, or (3) students may adjust their goals and modulate their efforts to fit teachers' classifications and expectations. Although the Pygmalion effect is a self-fulfilling prophecy in each of these cases, there are significant differences between them. Notice that the label influences students' performance *directly* only in the third case. In cases 1 (teachers' effort) and 2 (biased marks) the labels influence teachers, and teachers in turn influence performance or evaluation. This is a point to keep in mind for later, when we will discuss different types of reflexivity and interaction.

The Pygmalion effect is worrying: what if children's performance reflects teachers' expectations, rather than their true capacities and effort? Our view of education, and the way in which we grade and assess students, would have to be radically revised. Such worries are quite typical of reflexivity debates. Most discussions, especially in the early literature, are alarmistic. Merton claims that reflexivity is both common and malign in the social realm, for instance. One of his examples concerns racial discrimination: the members of a minority group ("negro workers," in Merton's old-fashioned writing) have a reputation

	J	B
I	2, 2	0, 0
E	0, 0	1, 1

FIGURE 9.1. Strikebreaking as an equilibrium.

for being opportunistic strikebreakers. As a consequence, they are excluded from trade unions. But because they are excluded from the unions, they have fewer opportunities to find a job. When employers look for workers during a strike, then, the members of the minority group cannot afford to decline the offer. So their reputation as strikebreakers is confirmed: they are strikebreakers because they are believed to be strikebreakers, and they are believed to be strikebreakers because they are strikebreakers.

The self-fulfilling prophecy is an equilibrium. Recall that in a Nash equilibrium the beliefs of the players are always consistent and correct. So the observation of behavior confirms the expectations of the players, and the expectations induce behavior that is consistent with the expectations. The equilibrium is supported by a causal feedback loop.

The incentives of the players are represented in Figure 9.1. The row player stands for the (white) members of the unions: assuming that they will go on strike in any case, the option they face is to either include (I) or exclude (E) African American workers from the unions. The column player represents African American workers: they can either join the strike (J) or break it (B). There are two possible equilibria: in principle all the workers would be better off if they went on strike together (IJ). But if Column has a reputation for being a strikebreaker, then Row's rational response is to exclude (E). Conversely, if Column believes that she is going to be excluded, she is better off breaking the strike. The outcome EB is an equilibrium. The strikebreaker label is a self-fulfilling prophecy, and strikebreaking becomes an attribute of "negro workers."

Notice that dependence on beliefs is not a peculiar characteristic of "bad" prophecies. By looking at the matrix we can see that good and bad equilibria are exactly symmetric: the outcome IJ is also dependent on beliefs, and beliefs and behavior would support each other in equilibrium. This is not a peculiar characteristic of Merton's example, for multiple equilibria are ubiquitous in social life. The impression that reflexivity is peculiar and malign comes from the fact that in many cases there is an alternative, which is feasible and better than the equilibrium in which the players are trapped.

The fact that there are many equilibria does not imply that they are all equally likely to occur though. That will depend on various factors, such as the configuration of individual preferences and beliefs, their distribution in the population, and the way in which people react to new information or small variations of behavior. Unfortunately the "negro workers" example is static, and cannot be used to show how these factors interact. We need a more complex, dynamic model to capture the dynamic aspects of feedback loops.

The dynamics of feedback loops is highlighted in Merton's second famous, and equally gloomy example. The story begins with a financially sound bank, with plenty of liquidity for its day-to-day business. Suddenly an unusual number of customers appear at its desks to withdraw their savings. To make sense of this unexpected event, a few other customers conjecture that the bank is in trouble, and for precaution decide to withdraw their money. This causes further queues and more withdrawals. Very soon the rumor becomes true—the bank is insolvent and goes bankrupt.

We can represent the underlying mechanism using a simple model of belief change (Figure 9.2). Let us assume that each depositor can choose only one of two actions: she can keep her money stored in the bank or withdraw. On the horizontal axis we represent the proportion of depositors who are expected to

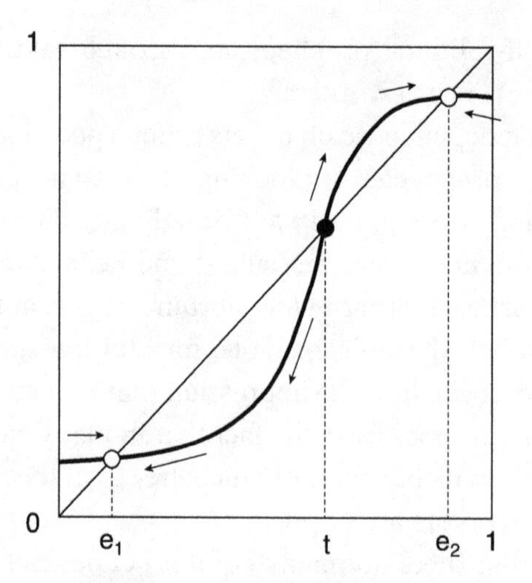

FIGURE 9.2. A Schelling's diagram with three equilibria.

withdraw, on a scale that goes from zero to one. For simplicity, we interpret this parameter as an aggregate function of the beliefs of all depositors, something like the average expectation of withdrawal in the population. On the vertical axis, we represent the proportion of depositors who actually withdraw (again on a zero-to-one scale). The diagonal line includes all the points where the expected number of withdrawals is equal to the actual number of withdrawals—that is, the set of all possible correct beliefs.

The S-curve in the diagram is a *propagation function*, because it represents how many people decide to withdraw, given their beliefs about withdrawals. If we suppose that customers observe each other's behavior in real time, we obtain a dynamic model known as a "Schelling diagram." The propagation function in Figure 9.2 crosses the diagonal in three different points: e_1, e_2, and t. These are self-fulfilling prophecies: people's beliefs about withdrawals (and, a fortiori, the soundness of the bank) are correct. These points are also equilibria, in the game-theoretic sense. Given her preferences and beliefs, no one has an incentive to change her behavior unilaterally.

With such a model we can analyze the forces that lead individuals to adjust their behavior as they gather more information and update their beliefs. The analysis shows that some outcomes are more likely to occur. As Merton pointed out, certain self-fulfilling prophecies are very predictable—indeed, they appear inexorable unless an external authority or unexpected event intervenes.

Let us start from the bottom-left tip of the S-curve. In this corner of the box we have those depositors who are going to withdraw their money no matter what happens. We can imagine these customers as people who need the money for some expense that had been planned independently of any turmoil in the financial sector. Even if no one else withdraws, they will do it. All the other depositors instead condition their choices on their expectations regarding the proportion of withdrawals. When the propagation function lies above the diagonal, more people are withdrawing (vertical axis) than it was expected (horizontal axis). There is a contradiction between beliefs and behavior that must be resolved. Once the queues begin to form in front of the bank, depositors process this new information and revise their beliefs accordingly. This corresponds to a shift to the right in the diagram: perhaps a few more customers will withdraw for precaution, but the tendency is to bridge the gap between expected and actual behavior. As long as we are above the diagonal, the same sequence of observation, belief revision, and withdrawal continues until the propagation function meets the forty-five-degree line.

When the reaction function lies below the diagonal, an adjustment in the opposite direction takes place. Fewer people withdraw their money than expected. Again, customers process this new information and revise their beliefs accordingly, causing a shift toward the left-hand side of the box. These two dynamics jointly imply that the adjustment process tends toward e_1 or e_2 (see the arrows in the figure). Notice that point t is also an equilibrium, because beliefs and behavior are everywhere consistent on the diagonal. But e_1 and e_2 are robust, in the sense that small deviations from these points trigger processes that bring the system back to its previous state. Point t in contrast is very fragile: it is a "tipping point," meaning that any small deviation will start a dynamics that will make the system converge toward e_1 or e_2.

Although in normal conditions the system will tend to gravitate around one of the robust equilibria, the tipping point may be crossed in anomalous circumstances. If this happens, a radical change of equilibrium will take place, as beliefs and behavior are quickly adjusted in the opposite direction. Social scientists call these dynamics *informational cascades*, and use them to explain abrupt institutional changes such as political revolutions or the collapse of the financial sector. When a tipping point is crossed the fate of the bank is sealed, unless a powerful authority (the government, or the central bank) intervenes to restore confidence in the institution.

How can it possibly happen? What triggers a run on a bank? Some event must push beliefs beyond the tipping point. If the system normally tends to gravitate around e_1, such an abrupt change can be brought about only by an external shock. It may be a new piece of information, for example the report of a rating agency saying that the bank has falsified its accounts. (Notice: the report need not be true—it is sufficient that it is believed to be true.) But a declaration of war, or a sudden political crisis may have the same effect. The common feature of such events is that everyone has to form very quickly new beliefs about the behavior of others, in unusual circumstances that they have never faced before. There is a lot of uncertainty, and people may go for the option that seems safest—to withdraw their savings for example—or simply copy the behavior of others. In these situations of panic, the whole system may crumble very quickly, and it may be difficult to intervene before it is too late.

The existence of multiple equilibria is well known to politicians and policy makers. The most difficult task for a revolutionary leader, for instance, is to convince his fellow citizens that a change of regime is about to take place. In Russia the attempted military coup of August 1991 failed when the national TV broadcasted unexpectedly the image of Boris Yeltsin standing on top of a tank in front of the house of parliament. Once the Russian people believed that the generals had failed to take power, their power was lost for good.

Or consider the behavior of the European Central Bank during the recent financial crisis. The possibility that the crisis could lead to the disintegration of the European Union was a major concern around 2011. The source of the problem was financial as much as political: the cost of servicing the public debt of countries like Spain, Italy, Portugal, and Greece would have escalated if investors had believed that the bonds would have been repaid in new devalued currency after an exit from the euro system. At the same time, it would have been politically costly for the leaders of other EU countries to promise that the debt would have been honored by the EU no matter what. It took months of difficult political negotiations before Mario Draghi, the new president of the European Central Bank, could state publicly that the ECB was ready to do "whatever it takes" to preserve the euro ("and believe me, it will be enough"). As things stand right now, it seems that simply saying it was more than enough.

In each of these examples there is a good equilibrium and its evil counterpart. The point e_1 is a nice equilibrium, for example, where the bank is strong and sound; e_2 is bad, because the bank does not exist. So the take-home message is that there is nothing intrinsically malign about reflexivity. Self-fulfilling prophecies may lead to equilibria that we like, as well as equilibria that we do not like (it depends on our preferences: a nostalgic of the Soviet Union would have preferred the 1991 coup to succeed for example). But without loops or self-fulfilling prophecies a lot of important institutions that we love and respect would certainly not exist. There would be no banks, for sure, but also no political leaders, no religious ceremonies, and no football teams.

The idea that our most cherished institutions are constituted by beliefs in equilibrium is quite disturbing. Merton clearly considered self-fulfilling prophecies perverse, for two reasons: because the equilibria that we like could be easily disrupted, and because new "bad" equilibria could be brought about merely by ideological propaganda or indoctrination. Fragility is clearly a concern, and we often try to conceal it by inventing theories that divert our attention away from reflexivity ("the European Union is historically *inevitable*," "African Americans are unreliable *by nature*"). But emphasizing the vulnerability rather than the robustness of institutions is like seeing the glass half empty or half full. Although the fragility of good equilibria is scary, the same fragility becomes an opportunity when the equilibrium is bad. Loops may be celebrated, as Douglas Hofstadter does in his best-selling book on *Gödel, Escher, Bach* (subtitle: *An Eternal Golden Braid*). They may have creative, generative power. They may bring new kinds and phenomena into being. And they may even bring about new ways of living—new human identities, as philosophers like Hacking and Michel Foucault have pointed out.

It took a long time for this simple truth to be appreciated by social scientists and philosophers. The equilibrium perspective on social institutions, which was developed independently and for other purposes, facilitates our understanding of reflexivity. What is a sound bank? In equilibrium, I do not withdraw my money because I believe that you will not withdraw, because I believe that you believe that I will not withdraw, and I believe that you believe that I believe that you will not withdraw, and so on. Social reality is constituted by beliefs about beliefs in equilibrium.

REFERENCES AND FURTHER READINGS

The concept "reflexivity" has a long history, but the term has spread in the philosophical and scientific literature only since the 1960s. Among its precursors, it is worth mentioning Morgenstern's self-defeating predictions (1928), the "Thomas theorem" (Thomas 1931), and Merton's (1948) self-fulfilling prophecies. The original "Pygmalion effect" was reported by Rosenthal and Jacobsen (1968). The term "reflexivity" as far as I know occurs for the first time in an article by Roger Buck that criticized Grünbaum's article on the self-fulfilling prophecy. In philosophy of science, the problem has been discussed by Grünbaum (1956), Nagel (1961), Buck (1963), Romanos (1973). The first application of equilibrium concepts to analyze reflexivity can be found in Grunberg and Modigliani (1954) and Simon (1954). Grunberg (1986) reviews some of this older literature. The simple model of informational cascades comes from Schelling (1978). Kuran (1995) uses the same model to analyze the processes that lead to political revolutions. The idea that reflexivity is not necessarily malign, but on the contrary may constitute the basis of all social reality was first put forward by Krishna (1971). Barnes (1983, 1989) generalizes this point, and highlights the tight relationship between forecasting and labeling. The recent renewed interest in feedback loops owes much to Hacking (1986, 1995, 1999), who, in turn, was deeply influenced by the philosophical historiography of Foucault (e.g., 1961, 1979). Mallon (2003) argues that Hacking's looping effects can have stabilizing as well as destabilizing effects, and makes an explicit connection with Lewis's theory of coordination equilibria. Finally, George Soros (1987, 2013), the financial guru, thinks that reflexivity "sets the natural and social sciences apart." Part of this chapter is based on an article devoted to Soros's theory (Guala 2013b).

CHAPTER 10

INTERACTION

Many philosophers and social scientists are *methodological pluralists*. Pluralists think that the social sciences must follow methods of investigation that differ from those of the natural sciences. Methodological pluralists are often ontological pluralists in the first place: they believe that differences in method are justified by differences in subject matter. Because social reality is radically different from natural reality, our methods must vary according to the domain of investigation.

There are many ways to be a pluralist, and some of them exploit reflexivity. The basic idea is simple: because the theories of social science are involved in reflexive loops, they influence their objects of investigation in a way that is precluded to natural science theories. Reflexivity demarcates social from natural science.

This idea goes back a long time. As we have seen Merton believed in demarcation in the 1940s, but abandoned it in the 1960s. The most recent attempt to revive demarcationism in the philosophy of science is due to Ian Hacking, who has introduced a new terminology and new conceptual tools. Hacking does not speak of reflexivity or self-fulfilling prophecies. His demarcation concerns *interactive* and *indifferent kinds:* human kinds are interactive, those of natural science are indifferent. Here is how Hacking illustrates the notion of interactive kind:

> We are especially concerned with classifications that, when known by people or by those around them, and put to work

> in institutions, change the ways in which individuals experience themselves—and may even lead people to evolve their feelings and behavior in part because they are so classified. Such kinds (of people and their behavior) are interactive kinds. (Hacking 1999: 104)

"Kind" is an ancient philosophical term. It was originally used by John Locke to translate the Aristotelian notion of genus (*genòs*). Centuries later it is still used in metaphysics and in philosophy of science, with different connotations. In general, it is used with a realistic slant: the world comes already sorted into kinds, before we look at it. The language of kinds is typically used in contrast with the language of sets or classes, collections of things that happen to be grouped for contingent reasons.

An influential conception—first articulated by John Stuart Mill (1843)—relates kinds with scientific inferences and properties. Consider the set of all white things: such entities, by definition, share only the property of being white. We cannot infer much else about them—other properties—from merely noticing that they are white. (*Something* can be inferred, of course: for example, white things reflect light and do not get warm in the sun; but not much can be said beyond that.) Other sets are more interesting, in contrast, because they support a lot of inductive inferences: from the fact that their members share some properties (they belong to the same kind) many other properties can be inferred. Take cats for example. Cats have nails, four legs, and a liver and eat mice. They dislike water, consume between twenty and thirty calories per pound of body weight each day, and have an average life expectancy of fifteen years. Classifying Missy as a cat gives us lots of information, because many properties can be inferred from its belonging to the kind *felis catus*.

An updated version of the Millian conception has been developed by Richard Boyd (1991). Boyd defines kinds as *homeostatic property clusters*. They are "clusters" because the properties of kinds tend to come in a package: they are highly correlated. Having one property makes it very likely that another property occurs too. They are "homeostatic" in two senses: (1) the properties are not clustered by chance—there are causal mechanisms that explain why they are correlated—and (2) the correlations are relatively stable. "Homeostatic" is a term borrowed from cybernetics, and refers to the tendency of a system to return to a normal state in spite of relatively minor perturbations. The term used in cybernetics to refer to the mechanisms that bring the system back to its normal state is "feedback loop" (notice how these terms recur—not by coincidence, for sure). The classic example of a homeostatic property cluster is a biological species, and the mechanism that ensures homeostaticity in that case is natural selection: whenever a new trait or property emerges by mutation, it tends either to disappear (if it decreases fitness) or to spread in the population, creating a new cluster (if it increases fitness).

Following Mill, let us call any class that has such properties a *real kind.* Real kinds are important for science because they support inductive inferences and generalizations. They are *projectable.* The properties and mechanisms of kinds are studied by the sciences, and scientific knowledge makes forecasting possible. If we know some properties, we can infer other properties. Scientific categories like "electron," "uranium," or "competitive price" refer to projectable kinds and real kinds.

Notice that real kinds, according to this conception, may overlap: the same entity may belong to different kinds. Some of these kinds may be natural in the sense of "studied by the natural sciences," and others may be social in the sense of "studied by the social sciences." For instance, I belong to the kind bipedal primate but also to the kinds male, father, husband, Juventus fan. All of these kinds may support inductive inferences and generalizations, so in principle it seems that one can be a realist about both natural and social kinds.

According to Hacking, however, social and natural kinds are not exactly on a par. There is a difference, and it has to do with the way in which we react to being classified into kinds. While natural entities are indifferent, the members of social kinds are not. They interact with classifications—they are interactive kinds.

> The classification "quark" is indifferent in the sense that calling a quark a quark makes no difference to the quark. (Hacking 1999: 105)

> A cardinal difference between the traditional natural and social sciences is that the classifications employed in the natural sciences are indifferent kinds, while those employed in the social sciences are mostly interactive kinds. The targets of the natural sciences are stationary. Because of looping effects, the targets of the social sciences are on the move. (Hacking 1999: 108)

Hacking thus proposes a demarcation criterion between the natural and the social sciences. The criterion has an ontological basis. It mirrors a distinction between two classes (kinds of kinds) with different features of some philosophical consequence. We will see later what these consequences may be. For the time being, it is clear that interactivity is related with reflexivity: social kinds are reflexive, natural kinds are not.

Hacking's demarcation criterion has not had an easy life. The problem is that interactive kinds are quite common, even in the natural realm. Mary Douglas (1986), who first pointed this out, mentioned microbes: calling an entity *staphylococcus* may have real consequences for its future existence (how it reproduces or dies out, for example, or whether it mutates into something else as a consequence of our attempts to exterminate it). The properties of *staphylococcus* may change dramatically because of the way it has been named. The same point has been made repeatedly by other critics, with a particular emphasis of animal and plant breeding: marijuana, dogs, and livestock are all interactive—they change their behavior as our categories change.

The consensus is that we cannot dodge these counterexamples and save Hacking's demarcation criterion. But although it does not capture a demarcation between natural and social science, the criterion captures a real difference between two kinds of kinds. There are many natural and social kinds that are not interactive in the strong sense, but there are also many that are, and this suffices to make the criterion interesting. In a substantial range of phenomena (from failing banks to discriminated minorities) the social sciences deal with interactive kinds after all.

Interactivity seems peculiar and interesting in itself, but does it have any significant implications? Are interactive kinds an obstacle to the application of standard scientific methods, for example? Are interactive kinds not real, in the Millian sense? Hacking claims that interactivity implies *dynamic nominalism*, the view that "the world does not come with a unique pre-packaged structure" (1999: 60). The accent is on "dynamic": unlike natural kinds, social kinds are "on the move." They change over time, and they do so (in part) because of our classifications. But what is the significance of this dynamic? Why does it matter, from an epistemological and ontological point of view?

On one reading, interactivity makes scientific inference difficult. Projectability (the capacity to support inductive inference) requires stability, but interactive kinds are transient. Since projectability is the distinctive feature of real kinds, interactive kinds cannot be real kinds in the Mill-Boyd sense. But the argument works only if interactivity implies instability, and this is doubtful, as we have seen in the previous chapter: some reflexive equilibria are very stable indeed. So what could the problem of interactivity be?

Hacking discusses interactive kinds extensively in his 1999 book, *The Social Construction of What?* Interactive kinds play an important role in his attempt to extract a philosophically interesting and defensible thesis from so-called social constructionism. In a nutshell, Hacking argues that many cases discussed by constructionists involve feedback loops, and that the social kinds that depend on these loops (interactive kinds) are contingent. They lack *necessity*, rather than stability.

Social constructionism is a notoriously complex and heterogeneous set of ideas. In the *Stanford Encyclopedia of Philosophy*, Ron Mallon identifies its core as follows:

> If there is any core idea of social constructionism, it is that some object or objects are caused or controlled by social or cultural factors rather than natural factors, and if there is any core motivation of such research, it is the aim of showing that such objects are or were under our control: they could be, or might have been, otherwise. (Mallon 2013: 1)

The first unifying principle of social constructionism thus is the rejection of naturalistic explanations of social kinds. The rejection of naturalistic explanations can be more or less interesting—and controversial—depending on its target. Take Italy (the nation) or Italian identity (the conscious feeling of belonging to an Italian nation or culture): few historians and scientists would deny that they are the product of a long and laborious process of social transformation. It took many centuries, from the poetry of Dante Alighieri to the political propaganda of Giuseppe Mazzini, to convince the people who live in the peninsula that they belong to a single community, that they are unified by a culture, and even that they should speak the same language. The alternative idea—that Italianhood is based on some set of biological traits—is ludicrous for a territory that in the course of the centuries has been invaded by Greeks, Romans, Goths, Mongols, Arabs, Normans, Spaniards, French, and Austrians, to name just a few, and whose inhabitants spoke mutually incomprehensible dialects until a century ago.

If Italianhood looks trivial, other cases are not. Social constructionism becomes controversial especially when it is applied to kinds that lie at the boundary between the social and the natural realm. Two classic cases are race and gender. The reason they are controversial is that race and gender are widely believed to be biologically determined, but at the same time are associated with specific social and economic roles. In many societies women, for example, are expected to raise children and to follow the orders of their husbands. African Americans are expected to excel in sports and music but not in philosophy or math. To say that these social kinds are socially constructed then is to say, controversially, that they are not based on natural (biological) traits.

To many, this may seem counterintuitive. Women should raise children because they are naturally suited for it—they are naturally bound to their kids, more caring, and disposed to make sacrifices for them. African Americans excel at sports because they are more athletic, taller, and stronger than Caucasian people. These intuitions are partly based on empirical evidence: if you live in a traditional society, for example, you will certainly observe that most women are better at child care than most men, if only because the latter have never been taught how to do it. But the intuitions are also based on a psychological propensity to find biological explanations more satisfying than social explanations. The psychologists who study this phenomenon use the expression *psychological essentialism* to refer to it.

Psychological essentialism is manifested in the tendency to posit the existence of necessary properties (essences) that are shared by all members of a kind. Possessing the XX chromosome, for example, is a good candidate for the essence of femalehood. Essentialism also invites us to exclude that an individual can belong to more than one social category at the same time (you must be either man or woman, black or white, German or French). It prompts to seek for superficial patterns of association in a population of individuals, and these patterns are interpreted as evidence of "groupness." Finally, it suggests that kinds and their boundaries are more rigid and less historically persistent than they actually are.

There is an obvious functional explanation of essentialism and of the tendency to naturalize social kinds. Recall Merton and the alarmistic literature on reflexivity: the idea that our most important institutions are rules in equilibria is worrying. It suggests that equilibria can be subverted, and new institutions created from scratch. So one way to protect our cherished institutions is to hide some alternatives from view. Only women can look after the kids, because men are not naturally gifted for that. Black kids must play basketball and white kids must play chess, for they are physically and mentally fit for these activities.

An important goal of social constructionism has been to challenge these intuitions, demonstrating that natural explanations of social roles are often unfounded. Some social constructionists have gone further, though, questioning the very idea of scientific explanation and embracing forms of relativism that have alienated many potential friends. This has been a mistake, because a pseudo-explanation is easier to displace if an alternative scientific explanation is offered in its place. Natural science is not the whole of science, after all, and social explanations of social roles are often better supported than biological ones.

The success of nonnaturalistic explanations varies from case to case, and this is not the place for a comprehensive assessment. My goal is just to show that the theory of institutions as rules in equilibrium makes us see the connection between constructionism, reflexivity, and interactive kinds more clearly. If an institution facilitates coordination in a game with multiple equilibria, then it is not inevitable—there is typically an alternative arrangement to the one that is currently in place. And this means that the institution is potentially open to critique and reform.

Interactive kinds are not inevitable because they are institutional kinds, and institutions solve coordination problems with multiple equilibria. Any given arrangement is a contingent effect of a contingent history. Our society could be organized differently. Consider the equilibria in Figure 10.1, for example: to make the

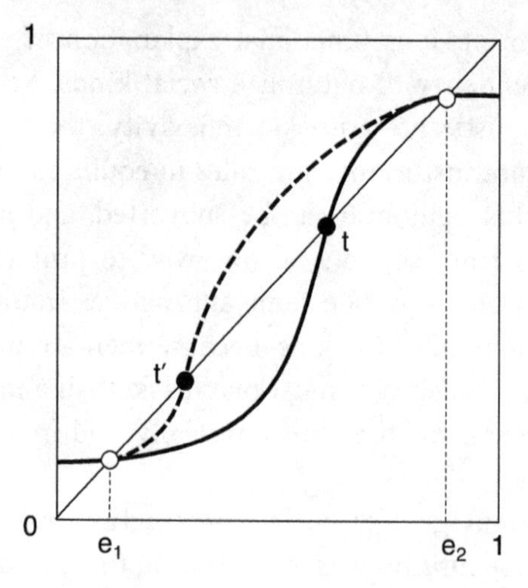

FIGURE 10.1. A Schelling's diagram with two reaction functions.

point more vividly, we can give them an interpretation in terms of new labels, predictions, and interactive kinds. Since homosexuality is one of Hacking's paradigmatic examples of interaction, I will use Schelling's model to tell a fictional story of sexual classification and gay liberation.

On the vertical axis, the model represents as usual the proportion of people in a population that chooses a given action. The action in this case is "coming out," declaring one's sexual identity publicly. Beliefs about the proportion of people who profess their sexuality in the open are represented on the horizontal axis. The S-shaped curves are propagation functions. The dotted function on the right has two "homeostatic" equilibria: in e_1 only a minority of people choose outing; in e_2 almost everybody comes out of the closet. Because preferences for outing are conditional on the number of people who are expected to do it, a group may be trapped in e_1. Even though most of them would prefer to come out, they do not do it because not enough people are expected to do it. Moreover, any small deviation from e_1 triggers a process of belief and behavior revision that brings the system back to the inferior equilibrium.

People are often discouraged from performing a certain action—like coming out—by its social consequences. Suppose for example that the sexual identity in question is commonly associated with mental illness, indecency, and immorality. If the term traditionally used to denote such an identity ("homosexuality") has this negative connotation, its use will generally discourage people from declaring their identity. The equilibrium e_1 thus is very stable, but by no means necessary. Some people—all of them, perhaps—may be aware of this contingency, and envisage a different equilibrium where everybody profess their sexuality openly. The problem is how to get out of the trap, crossing the tipping point that would trigger an informational cascade.

One way to do it is to change people's incentives, manipulating the sanctions that discourage them from coming out (the "normative apparatus" that disciplines homosexuality, to use a Foucauldian expression). The task is daunting and the story is complex, of course, but here is a stylized account: suppose that a new term ("gay person") is introduced, to express a new conception of the same sexual orientation. Suppose that a number of people who occupy prominent roles in society (artists, designers, and musicians) begin to self-identify as gay. If "gay" connotes a positive, legitimate, interesting way of life rather than a moral defect or medical condition, people's preferences and propagation functions may change significantly. In Figure 10.1 the propagation curve shifts upward and to the left. As a consequence, the tipping point (t') also shifts to the left. This means that a small deviation from e_1 may now trigger an informational cascade. A minority of activists could for example cross the tipping point, persuading others to follow. As more people come out of the closet, the system converges to e_2. The new label changes beliefs about classified people, beliefs change behavior, and so forth, until a new identity is stabilized in the superior equilibrium.

Although the story is stylized, it captures a general mechanism. The point is that in equilibrium a label and the behavior it describes may be extremely stable, without being inevitable. A person who is classified as "gay" may engage in a number of predictable activities and may possess a variety of correlated behavioral traits, partly because she sees herself as gay. A gay person may vote for a certain political party, listen to a certain kind of music, or dress in a certain way. These correlations may all be used to issue predictions. In fact they are routinely used by political pollsters, social scientists, and marketing experts to anticipate the behavior of people who are labeled in a certain way.

And this is unsurprising, in a sense: an important function of classifications is to stabilize behavior, facilitate convergence on equilibria, generate behavioral regularities. So ordinary labels and folk classifications offer social scientists the opportunity to come up with generalizations that support inductive inference. Institutional terms refer to behavioral propensities, in part, because the actors themselves use them as coordination devices. The equilibrium approach explains why a social category or institution can be at the same time contingent (because it is one among many possible equilibria) and stable (because it is an equilibrium).

Social constructionists should welcome this analysis. It explains why their project is legitimate and feasible (it can succeed), and why challenging everyday classifications is an important part of the constructionist task. It also explains why it may be very difficult to succeed, and why some classifications may be more entrenched than others (the equilibria are more robust). However, it also shows that the antiscientific attitude of much traditional constructionist literature is misplaced. It is possible to challenge biological explanations of social roles without challenging scientific explanation tout court.

It is true that transitions across equilibria are usually difficult to predict, because the propagation function is unobservable. But this does not make the dynamics of reflexivity inexplicable from a scientific point of view. It is also true that an adequate explanation will sometimes have to account for the fact that transitions are triggered by changes in people's normative expectations. This aspect was emphasized in chapter 5: by charging behavioral rules with normative force, it is possible to change individual incentives and create new equilibria that did not exist before.

Paul Griffiths (2004) has proposed to use the term *normative kind* to denote those entities, properties, and phenomena that are at the same time the subject of descriptive investigation and moral evaluation in a given community. Typical examples are emotions, sexual orientations, and mental illnesses. The way in which these kinds are described is sensitive both to epistemic and normative considerations.

> Normative kinds are open ended because normative projects as well as epistemic projects provide motives for altering extension and intension. Just as a category can be altered for the sake of inductive power, a category can be altered, for example, as part of a project of social reform. (Griffiths 2004: 908)

The introduction of a different terminology to denote the sexual identity of "gay" people is a case in point. Griffiths points out correctly that such morally sensitive kinds can be studied scientifically (epistemic project) and simultaneously may be the target of political critique (normative project). Sometimes the epistemic and the normative projects proceed in parallel, unproblematically, and sometimes they pull in opposite directions. The influence is reciprocal: changes of classification prompted by normative concerns may influence changes in descriptive classification, and vice versa. Later in the book we will see an example of this complex dynamic, when we will examine the case of same-sex marriage.

For the time being, let me conclude the chapter in an ecumenical mood: contingent social kinds offer opportunities for social reform. Social reform is usually driven by normative concerns, but at the same time must be sensitive to the reflexive mechanisms that make current classifications resilient and robust to manipulation. That is why normative projects and epistemic projects go hand in hand, and why we should better accept that normative kinds are no less real than other scientific kinds.

REFERENCES AND FURTHER READINGS

The most popular philosophical term to distinguish genuine from bogus kinds is "natural kind," but for obvious reasons it is preferable not to use it in a book like this. Mill's "real kind" is better suited, and I will stick to it from now on. Hacking has written extensively on the tradition of real kinds; see especially Hacking (1991, 2007b). The philosophical approach that links real kinds with inductive inference goes back to Mill (1843), Whewell (1847), and Venn (1866). "Projectable" is a term introduced by Goodman (1954). The modern revival of kinds is due especially to Putnam (1975) and Kripke (1972), to whom we owe the unfortunate association with essentialism—the idea that every entity has a set of properties that determine necessarily what sort of thing it is. Pluralism about kindhood is generally hostile to essentialism; see, for example, Dupré (1993) and Khalidi (2013). The homeostatic property cluster theory is articulated in Boyd (1989, 1991) and discussed critically by Wilson (1999). There are many critiques of Hacking's distinction between interactive and indifferent kinds: see in particular Douglas (1986), Bogen (1988), Haslanger (1995), Cooper (2004), Khalidi (2010). Mallon (2003) first pointed out that interactivity may be a source of stability, and Kuorikoski and Pöyhönen (2013) have outlined a causal-mechanistic interpretation of feedback loops. Recently Hacking has become pessimistic about the demarcation criterion based on interactivity: he claims that although interaction is a real phenomenon, it does not demarcate sharply the natural from the social realm (see Hacking 2007a). On psychological essentialism, see, for example, Gelman (2003), and Rothbart and Taylor (1992) for an application to social kinds. Machery (2014) discusses the relevance of essentialism in the context of social ontology. Finally, the story of the term "gay" has been drastically simplified in the text; over time the term has been charged with negative as well as positive connotations, and there are signs that the pendulum is swinging back to the negative side again. The literature on the history of (the concept of) homosexuality is enormous, but Pickett (2011) offers a brief philosophical introduction.

CHAPTER 11

DEPENDENCE

Real (adj.):

1. Actually existing as a thing or occurring in fact; not imagined or supposed.
2. (Of a thing) not imitation or artificial; genuine. (*Oxford English Dictionary*)

"Real" is a polysemic term. When we speak of *real kinds* we may be saying at least two different things: (1) that they exist independently of our theories, beliefs, representations, or (2) that they are genuine kinds. When we say that hippogriffs are not real, for example, we are using the term in the first sense. We are using it in the second sense, in contrast, when we say that Tiger Woods is a real champion. In the tradition that goes back to John Stuart Mill, and that was at center stage in the previous chapters, the expression "real kind" is interpreted in the second manner. A genuine kind supports inductive inferences and generalizations. A nongenuine kind does not—it is a bunch of things that we group together for one reason or another, but *not* because of their correlated properties.

The two senses of "real" however are related. An important and peculiar feature of many social categories is that they are involved in reflexive loops. As a consequence, it seems that these social kinds cannot exist independently of our thoughts and representations. They fail to be real in the sense of mind-independent. Most philosophical formulations of realism in fact include mind-independence as an essential requirement. According to Richard Boyd, for example, the doctrine of scientific realism holds that "the reality which scientific theories describe is largely independent of our thoughts or theoretical commitments" (1983: 45).

But what does it mean to say that real kinds must be independent from our representations? Two interpretations are possible: the first one takes it as a requirement of *causal* independence; the second one as *constitutive* or *ontological* independence. Hacking occasionally speaks of "a person being constituted as a certain type of being" (2002: 24), but most of his examples of theory-dependence have an unmistakable causal flavor. To attribute a constitutive interpretation to him would require a considerable stretch of imagination. The definition of reflexivity that I have given in chapter 9 in fact was based on the notion of causal loop (X causes Y, and Y causes X), and game-theoretic equilibria are also supported by loops of this sort: there are causal mechanisms ensuring that beliefs and behaviors sustain each other. Small deviations will trigger a "homeostatic" mechanism that will bring the system back to its equilibrium state.

In chapter 10 I have argued that causal dependence on representation does not undermine the reality of social kinds. Reflexive relations do not prevent social scientists from doing the things that scientists do. Explanation and prediction, in particular, are not made impossible by feedback loops of the Merton or Hacking type. Reflexive loops sometimes do take the form of self-defeating prophecies and do disrupt inference, to be sure. But they often facilitate inference, as in self-fulfilling prophecies. So causal dependence on representation is not an intrinsic obstacle to the existence of real social kinds.

How about noncausal dependence though? Dependence comes in many different forms, but for the purposes of this chapter we need only a rough definition of the relation that philosophers have variously called "existential," "ontological," or "constitutive" dependence. According to a standard characterization,

$$X \text{ depends ontologically on } Y = \text{necessarily, } X \rightarrow Y.$$

For example, if Francesco's existence depends on the existence of Francesco's body, then necessarily whenever I exist my body also exists. There are subtle counterexamples to this rough definition, so we will take it only as a first approximation. But it is a useful approximation, if it can be used to clarify how social entities may be ontologically dependent on mental states. To begin with, ontological dependence is asymmetric, in the sense that apart from peculiar cases (like the dependence of an entity on itself) we can usually assume that X's dependence on Y does not imply Y's dependence on X. (The existence of my body, to use the previous example, does not necessarily depend on my existence.) Moreover, ontological dependence and constitution are the converse of one another: if X constitutes Y, then Y depends ontologically on X.

The dependence that is of interest to us is the dependence of social objects on mental states. There is a sense in which this may be true by definition: mental states like beliefs and desires are often used to distinguish human action from other types of behavior (automatic behavior, for example), so any social or collective action may be dependent on mental states in this sense. This is not, however, the sense in which the thesis of ontological dependence is usually interpreted: the idea is rather that the existence of social objects is dependent on the way in which *the objects* themselves are represented. For example, suppose that the mental states are propositional attitudes like beliefs or expectations. Let us also suppose that the mental states are shared in a given community—they are *collective attitudes* (CA). The dependence thesis then states that

$$\text{necessarily, } X \text{ is } K \rightarrow CA(X \text{ is } K),$$

or, in other words, if a token entity X belongs to the social kind K, then necessarily the individuals in the relevant social group collectively believe that X is K. The existence of money would depend on mental states, for example, in the sense that for a piece of paper to be money it would be necessary that it is represented as money in the relevant community.

Notice that social entities are not *just* representations: a dollar bill for example is a piece of paper, the hall of parliament is a building, and the chancellor of my university is a fifty-year-old professor of medicine. The important point—the feature that makes these entities *social*—is the fact that many people hold certain beliefs about them: the piece of paper is believed to be a dollar bill, the building is believed to be the hall of parliament, and so on and so forth. Analogous cases that have been widely discussed by metaphysicians are statues and works of art. The piece of marble sculpted by Michelangelo would not be a statue of David if there were no people (artists, art lovers, critics, historians) who believe that it is a statue. The *David* is constituted by a piece of marble *and* by these attitudes. So it may be argued that social entities and works of art both exist materially and are dependent on our representations. (I am not saying that this is a correct theory of art, but it is a view that has been defended and that helps clarifying the relation of dependence that may be relevant to social entities.)

Making social kinds dependent on representation in this way has some interesting philosophical implications. First, it seems to imply antirealism in the Millian sense. If dependence is understood in a noncausal fashion, in fact, it is not clear how ontologically dependent kinds may have the properties that real kinds are supposed to have. Recall that according to the Millian view real kinds must first and foremost support inductive inferences. But the correlations between properties that make inferences and generalizations possible hold in virtue of contingent facts, in particular the causal mechanisms that, according to Boyd (1991), ensure the "homeostatic" stability of the property clusters.

If the dependence view is interpreted *causally*, we can tell a plausible story about the way in which the creation and maintenance of behavioral regularities is dependent on human representations. But under the noncausal or ontological interpretation it is not clear how these correlations can be ensured by means of representation. The dependence view thus drives a wedge between the criteria for social kindhood (dependence on representation) and those for real kindhood in the Mill-Boyd sense (projectability). The conclusion that social kinds are not real kinds becomes compelling: they are unified conceptually, by our classifications, rather than by causal mechanisms that guarantee the co-occurrence of their properties. They are unlikely to be projectable and cannot be used for scientific purposes.

This conclusion follows from the adoption of a simple thesis that, following Boyd, I will call the *principle of the metaphysical innocence of representation*:

> theories and classificatory principles are *metaphysically innocent*: they affect the causal structure of the world only via the operation of intermediary causal mechanisms which supervene on the causal structures studied by the various special sciences and not also in some additional way studied only by philosophers practicing conceptual analysis. (Boyd 1991: 144–45)

I agree with Boyd that this principle does not require an explicit defense, because "it appears to be underwritten by fundamental conceptions of causal relations common to all the established sciences" (1991: 145). The burden of proof lies with those who intend to deny the metaphysical innocence of representation; minimally, these philosophers owe us an explanation of how contingent properties and causal relations can be fixed by our representations in a noncausal way. As far as I know, no one has attempted to give such an explanation. Most philosophers who endorse a noncausal version of the dependence thesis are inclined to retain the principle of metaphysical innocence and to give up realism about social kinds.

But, surprisingly, giving up realism may have some advantages on a different front. From an epistemic point of view, mind-dependent entities seem to be in some way more transparent and accessible than mind-independent ones. For example, if a piece of marble is a statue, arguably the art world (critics, historians, collectors) must necessarily be aware that it is a statue, since being a work of art is partly constituted by the art world's belief that it is a work of art. And the same applies to a banknote: people cannot fail to see that a piece of paper is money, if money is constituted in part by people's belief that a certain entity is money. Considerations of this type invite the formulation of a general philosophical thesis that I will call *infallibilism about social kinds*: we cannot be wrong about the nature of social kinds that depend ontologically on our representations.

Infallibilism and antirealism go hand in hand because their opposites—realism and fallibilism—are tightly related. If the objects that we represent are independent of our representations (realism), then they can be misrepresented (fallibilism); if they are dependent on representations (antirealism) then they cannot be misrepresented (infallibilism). Amie Thomasson has made the connection explicit, articulating realism by means of three principles:

> *Extensionality*: there is a kind with natural boundaries that determine the extension of the term independently of anyone's concept(s) regarding the kind.
>
> *Error principle*: since these boundaries are not determined by human beliefs about those boundaries, any beliefs (or principles accepted) regarding the nature of Ks could turn out to be massively wrong.
>
> *Ignorance principle*: for all conditions determining the nature of the kind K, it is possible that these remain unknown to everyone. (Thomasson 2003: 583)

According to Thomasson none of these principles hold in the case of social kinds. Extensionality is contradicted by the idea that social entities are ontologically dependent on our representations. Violations of the Error and the Ignorance principles follow quite naturally. As David-Hillel Ruben has put it,

> The essential point of realism . . . is that it is always possible that our theories are wrong: error and mistake are always possible. Where there is this distinction between theory and theorized reality, our theory, or our beliefs about that reality more generally, may have failed to grasp adequately or comprehend what they are about. (Ruben 1989: 60)

In the social realm, in contrast, "widespread general classificatory belief that there are things of social kind *s* is sufficient for there being things of social kinds *s*." As a consequence, "consistent and widespread error regarding general classificatory beliefs about society is indistinguishable from reality" (Ruben 1989: 74). Both Ruben and Thomasson defend infallibilism about social kinds. If ontological dependence on representation implies denial of the extensionality principle, and the latter implies infallibilism, then infallibilism about social kinds is implied by the dependence thesis.

Ontological dependence thus would carry both good and bad news. The bad news would be ontological, because social kinds could not be real kinds; and the good news would be epistemological, because we would have direct epistemic access to the social world. Although antirealism and infallibilism form an odd couple, philosophers have never been shy of defending strange doctrines, and in fact the odd couple has attracted plenty of followers over the years. Many philosophers take both the good (infallibilism) and the bad news (antirealism) at face value. They argue simultaneously (1) that knowledge of the social world can be attained directly, without the risk of error; and (2) that this knowledge cannot be used to make inferences like those of natural science. Antirealism and infallibilism are ubiquitous in the *Verstehen* and hermeneutic tradition that runs from Wilhelm

Dilthey to Georg Gadamer, Paul Ricoeur, and Charles Taylor. But they are also common in the Wittgensteinean tradition, for example in Peter Winch's influential critique of the naturalistic pretensions of social science and in Michael Dummett's philosophy of mathematics. Versions of thesis 1 usually say that those who share a common culture (a common language and common practices) have a direct and intuitive understanding of social categories. This understanding comes from mastering the relevant social practices—a "way of life"—and is precluded to outsiders. Thesis 2 usually says that the intuitive understanding of the insider is the *only* form of understanding that can be attained in social science. Social phenomena cannot be studied empirically as realities that are independent of our thoughts. Indeed, taking the detached point of view of the outsider—as empiricist social scientists try to do—guarantees that understanding will never be attained. According to Taylor, for example, classic scientific criteria like predictive accuracy are not a good test for social theories:

> We cannot measure such sciences against the requirements of a science of verification: we cannot judge them by their predictive capacity. We have to accept that they are founded on intuitions that all do not share, and what is worse that these intuitions are closely bound up with our fundamental options. (Taylor 1971: 51)

What shall we make of claims like these? Infallibilism and antirealism follow from the thesis of ontological dependence, and so far we have not been given a good reason to take the latter on board. The dependence thesis as far as we know may well be false. Eventually I will argue that it is false, and that we can happily be realists and fallibilists about social kinds. But in order to secure that conclusion some preliminary work is required, mainly to put infallibilism and antirealism in a shape that cannot be too easily refuted. Since Ruben and Thomasson have done most of the job already, in the remaining part of this chapter I will follow their lead. The critical part will have to wait until the next chapter.

First we must get rid of some cheap counterexamples. Clearly infallibilism does not hold for *all* social kinds. As Ruben points out, the existence of some social entities does not require explicit recognition by the members of a community. "Exploitation, alienation and many other social phenomena can exist undetected" (1989: 74). According to Thomasson (2003: 606), similarly, infallibilism does not hold for the phenomena that occur as unintended consequences of a series of actions; and for the causal relations that hold among social entities and kinds.

Notice that Ruben's and Thomasson's infallibilism would still leave a lot of work for traditional (empiricist) social science to do. But the infallibilist thesis would remain true for a significant subset of social facts, and knowledge of these facts would be based on epistemic foundations that are different from those of standard scientific knowledge of the empiricist type.

It would be useful at this point to have a general criterion to identify the class of entities for the existence of which a collective belief (representation) is necessary and sufficient. This would give us a more precise characterization of the class of things for which traditional realism fails. Thomasson uses Searle's constitutive rule theory as a criterion. In *The Construction of Social Reality* Searle distinguishes between institutional and merely social facts. The distinction can be cashed out in terms of the acceptance of a constitutive rule. The existence of an institutional fact requires collective recognition that something (X) is an institutional entity (Y) in a certain context. A social fact in contrast needs to involve only shared beliefs about actions. For example, if all the members of the Corleone family jointly believe and desire that a member of the Provenzano family should be disposed of, when they plan and execute the killing they are performing a social action, and the ensuing feud is a social fact. The existence of social kinds (and the derivative facts) depends on a human activity of creation and maintenance of such kinds and facts; representations (beliefs, goals, intentions) are part of this activity, but only in a causal way. The murder is not part of a feud, in particular, in virtue of the fact that the Corleones and the Provenzanos *represent it as* a feud.

Now consider the special class of institutional facts. An institutional fact according to Searle exists in virtue of one or more constitutive rules of the form "X is Y if C," where X is a preinstitutional entity, Y is an institutional term, and C is a set of conditions that X must satisfy in order to be Y. For example, Vito Corleone is a convict because he has been sentenced to ten years in prison by a jury at the end of a fair trial. Here it seems uncontroversial that *someone* must represent the constitutive rule for Vito Corleone to be a convict. If no one believes that the individuals who have been sentenced by a judge are convicts, then there are no convicts.

This is in line with some examples of ontological dependence that I have mentioned earlier. There is a sense in which Michelangelo's *David* would not be a statue if nobody thought of it as a statue. Duchamp's *Fountain*, similarly, would be just a urinal if nobody thought of it as a piece of art. The artists and art lovers whose representations constitute the urinal as a piece of art cannot be in error when they identify Duchamp's *Fountain* as an artwork, because their judgment is what makes the urinal a work of art.

Giving up the extension principle seems to be a direct consequence of the constitutive rules theory. The boundaries of an institutional kind must depend on people's acceptance of a constitutive rule for it to be an institutional kind at all. The rejection of the Error and Ignorance principles seems to follow rather unproblematically, once you go down this route:

> Our acceptance of a set of conditions C as sufficient for being K is constitutive of what conditions suffice for being K, so what conditions there are is constituted by what conditions we accept. As a result, we could not turn out

> to be mistaken—our acceptance of the set of conditions C declaratively establishes the conditions for being K rather than attempting to describe preexisting and independent conditions for being K. So the Error Principle fails: any conditions we accept as sufficient for the existence of Ks must be free from error. (Thomasson 2003: 588–89)

Infallibilists' challenge to realism, if successful, would have deep implications regarding the methods of investigation that are appropriate for social ontology. If infallibilism is correct, then some general facts concerning social reality can be known a priori. A competent member of a community needs only examine her own beliefs and linguistic practices, in order to find out what it means for something to be K. The judgment of the competent member would be infallible, and this would distinguish sharply (some) investigations of the social world from investigations of natural reality. While empirical research would be always necessary to generate knowledge about the natural world, armchair conceptual analysis would maintain an important role in the realm of the social.

For the sake of clarity, it is important to distinguish infallibilism from the (spurious) claim that we cannot be wrong about *specific instances* of social classification. Infallibilists do recognize that the satisfaction of conditions C in the formula (X is an instance of K if C) is an empirical matter, and hence that individual acts of classification can go wrong in various ways. Specific issues of institutional classification therefore cannot be resolved a priori. Whether a token X counts as K or not is a contingent fact, for the possession of C is itself a matter of fact to be established empirically. Infallibilists however claim that the proposition "X is K if it has properties C" is true a priori. That a piece of paper counts as genuine money in virtue of the fact that it has been issued by the Bank of England is a fact that does not require empirical research in the same way as the fact that water is H_2O. It holds by mere stipulation (or collective agreement) by the members of the relevant community. Infallibilism is a thesis about our knowledge of general facts concerning the nature of social kinds. It concerns our knowledge of what it is, or what it means for something to be a thing of type K, and does not preclude errors in identifying token entities as K-things.

Another important proviso is that infallibilism cannot sensibly require that *every* member of a social group explicitly accepts the conditions C that make X an instance of K. This would clearly be unrealistic: lay people for example do not know exactly what conditions must be satisfied for a certain individual (Elizabeth Mary Windsor) to be the queen of England. Not only most people have no idea what the relevant conditions exactly are in cases like this; they often have false beliefs about them. (They tend to identify "blue blood," descent, and similar features as sufficient for example—whereas in fact none of them is.)

Another related problem is that beliefs about institutions are usually heterogeneous in a population—especially if the population is large. Cognitive scientists and anthropologists have noticed that the cultural transmission of representations is usually imprecise and unfaithful to the original, so that as a consequence there is rarely a single concept or representation that is collectively accepted by all the members of a group. The metaphor of a group of people sharing a belief with the same content ("we believe that Elizabeth is the queen in virtue of being a non-Catholic, legitimate descendant of King George, etc.") is a misleading idealization at best, and we should not take it literally.

So how can we salvage the infallibilist thesis that some social facts can be known a priori? One option is simply to shrink the size of the group, or more precisely of those members who are entitled to know on behalf of the group. Raimo Tuomela has introduced for this purpose a distinction between "operative" and "non-operative" members of a social group:

> It must be emphasized that in the case of developed societies there is division of labor also with respect to collective

> acceptance. Thus not all members of the community actually need to know all the details and may not even have heard of the [institution] in question [for that institution to exist]. (2002a: 200)

> In such collectives it is the operative members for decision who decide what will be money, for instance. In this realistic cases the other, non-operative members only need to tacitly accept what the operative members have decided. (2002b: 427)

This move preserves a principled distinction between correct and incorrect applications of a concept (or correct and incorrect attributions of kind membership). The distinction between, say, a legitimate and an illegitimate sovereign may be independent of lay people's beliefs about the institutional conditions C that apply in this case. All they have to (tacitly) agree upon is that there is a procedure for resolving such matters, and that certain experts know how to resolve it. Once this is agreed, the "ontological consequences" of this fundamental collective acceptance trickle down, so to speak, to whatever logically follows from it.

In such cases knowledge of institutional kinds is to be interpreted as "social knowledge" or "expert knowledge," knowledge that is stored somewhere in society, perhaps readily accessible only to a minority of experts. This is not a far-fetched suggestion: we speak of knowledge in this sense when we say that "Great Britain knows how to build the atomic bomb," even though the vast majority of British citizens have no idea how it can be done; or when we say that "twenty-first-century mathematicians know how to demonstrate Fermat's Theorem," even though perhaps only a handful of people are able to understand and reproduce the proof. Under this interpretation what some individuals (even the majority of individuals) believe is quite irrelevant—they can be systematically wrong and yet infallibilism may be true, because *as a society* we cannot be wrong about institutional kinds.

Even experts' acceptance, however, should not always be taken literally. The nature of social entities is certainly something we do not pay much attention to, for most purposes, in our everyday lives. We tend to deal with social kinds thoughtlessly, and we appreciate the conventional nature of social reality only occasionally, upon reflection. To account for this fact most collective acceptance theorists give up the requirement that collective acceptance must be constantly operative. Searle has famously introduced the concept of the "Background"—a set of mechanisms and dispositions that subconsciously and automatically support our social practices without requiring full cognitive engagement with the logical presuppositions of such practices. Similarly, Tuomela posits the existence of "virtual" mechanisms that would bring individuals' behavior back in line, were certain disturbing factors to disrupt the regularity of a social practice:

> The we-attitudes need to be respected in the various institutional activities undertaken in the institution, but, being dispositional states, they need not be made occurrent and reflected upon in normal circumstances but only in cases of institutional breakdown (or something analogous). (Tuomela 2002b: 426)

Such a "dispositional" view preserves a non-vacuous meaning for the denial of the Error principle that is a central plank of the infallibilist position. Even though the relevant actors are not constantly entertaining the thought that X is K if it satisfies conditions C, they can nevertheless retrieve such thought in the appropriate circumstances.

This is the final proviso that we need to shield the infallibilist thesis from cheap counterexamples. Again, let us notice that the domain of infallibilism has shrunk considerably, and that this elaborate version of the thesis cannot be used to defend the strong claims that some hermeneuticists and Wittgensteinean and methodological pluralists hold dear. In particular, it falls short of demonstrating that the social sciences in general ought to follow different methods from those of the natural sciences, because a wide variety of social phenomena would still have to be studied empirically. Still, the amendments are not so strong to make infallibilism trivial—a non-vacuous infallibilist thesis is preserved. Even though not all members of a group know that X is K in virtue of having C, some of them—the "experts"—do. And even though perhaps no one is always consciously aware of its institutional character, those who are in charge of determining and recognizing K-ness are able to form the appropriate mental states when required. We are now ready to challenge this version of infallibilism, debunking the thesis of ontological dependence upon which it is based. The next chapter will be entirely devoted to this task.

REFERENCES AND FURTHER READINGS

On the multiple meanings of "real," see Austin (1962, chap. 7) and Hacking (1983: 32–33). The concept of dependence has generated a sizable literature (see, e.g., Correia 2008; Tahko and Lowe 2015). Unfortunately metaphysicians have typically discussed types of dependence—like the dependence of a whole on its part—that are not particularly useful for the issues that we are interested in. A notable exception is Thomasson's work on fictional objects (1999, chap. 2) and Epstein's (2015) book on individualism. Colingwood (1946), Dray (1957), Winch (1958), and Taylor (1971) are classic statements of the hermeneutic position in Anglo-American philosophy. But hermeneutics is a distinctively European affair, so one may want to look also at Gadamer (1960), Ricoeur (1965), Habermas (1968). The antirealist implications of ontological dependence are sketched briefly in Boyd (1991), but on the distinction between causal and noncausal dependence see also Kukla (2000) and Mäki (2012). Ruben's (1989) and Thomasson's (2003) infallibilism is discussed in Guala (2010), upon which a large portion of this chapter is based. To define a sensible version of infallibilism about social kinds, I have relied extensively on Searle's and Tuomela's work. The dependence formulas that I use, in particular, are almost identical to the formula for reflexivity introduced by Tuomela (2002a: 140). But it is important to stress that Tuomela rejects infallibilism explicitly; Searle (1995) is more ambiguous, as we shall see in the next chapter. I owe Olivier Morin the point on the heterogeneity of beliefs in a population; on the imperfect transmission of representations, see Sperber (1996), and on the tricky metaphor of sharing the same content, see Turner (2010, esp. chaps. 5 and 6).

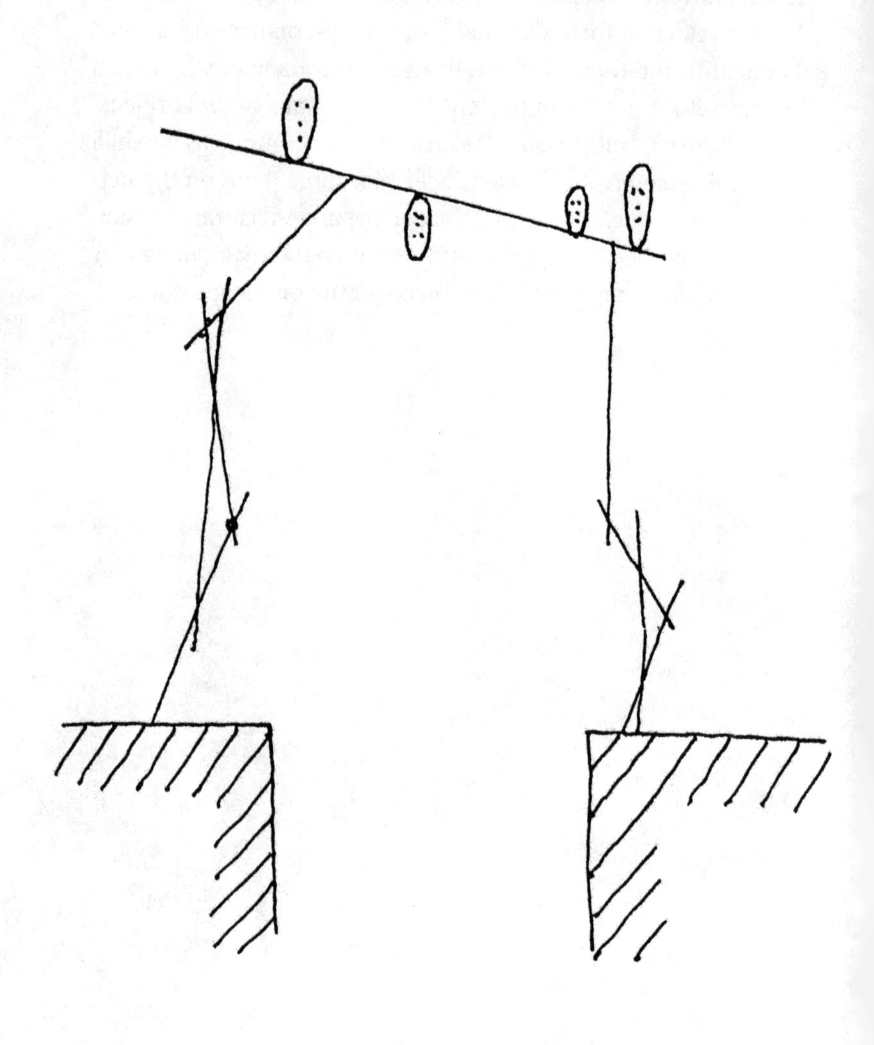

CHAPTER 12
REALISM

According to the dependence thesis, institutional entities depend for their existence on our representations. When dependence is interpreted causally, the thesis is not a threat to the reality of institutional kinds. When dependence is interpreted noncausally, in contrast, the thesis seems to imply antirealism and infallibilism. We have seen that many philosophers have endorsed this odd couple. But antirealism and infallibilism hinge on the dependence thesis. If dependence falls, they fall as well.

In this chapter I will argue that the dependence thesis is false. For the argument to take off, it will be necessary first and foremost to give a precise definition of the dependence thesis. As a first approximation, in the last chapter I proposed the following formula:

$$\text{necessarily, X is K} \longrightarrow \text{CA(X is K)}.$$

In this formula the arrow symbol stands for the relation of ontological dependence. X is a token entity that is a member of kind K. CA is a propositional attitude like collective acceptance, belief, or recognition. Muhammad Ali Khalidi (2015) has pointed out that the propositional attitude may be directed toward two different targets at least. In the formula above it is aimed directly at an individual. For example, if we all believe that Mick Jagger is the leader of the Rolling Stones, then he is the band's leader. But in other cases the attitude is directed toward the kind itself, or, more precisely, toward some conditions that are deemed sufficient for an entity to belong to K. In the latter case, in other words, people accept a general theory of what being a band leader amounts to. For instance, they may believe that if one of the members writes all the songs of the band then he or she is the leader. In this case, the attitude is directed toward the kind itself.

Ontological dependence, in the case of institutional entities, involves attitudes of this sort. More formally, kinds are constituted by the acceptance of a set of conditions that make an individual X a member of K, and by the satisfaction of these conditions:

$$\text{necessarily, X is K} \rightarrow [\text{CA(X is K if C) \& C}].$$

Notice that this formula includes a constitutive rule à la Searle (X is K if C). The existence of K depends on the acceptance of the constitutive rule, and the attribution of K-hood to a specific entity X depends also on the instantiation of the conditions C.

The first, shorter formula arguably depends on the second one, because implicitly or explicitly the members of a community must apply criteria to decide which tokens are to be accepted as members of K and which ones are not. By stating the conditions C, the second formula simply makes the criteria explicit. As an example, take the paradigmatic case of money: a particular piece of paper is money if and only if we accept that in order to count as money a paper bill must be issued by the central bank, and this bill has been issued by the central bank. To simplify the discussion I will refer primarily to the second version of the formula from now on. This should have no significant consequence, because everything that I will say will be applicable to the other version as well.

The dependence formula implies infallibilism about institutional kinds. It says that if X is an institutional kind then necessarily we know the conditions that ought to be satisfied for X to belong to K. If we read "X is K if C" as a theoretical statement, then the formula says that the existence of a kind implies necessarily the representation of a theory of that kind. And the theory moreover is necessarily true: if K exists we cannot be wrong about the nature of K (its "kindhood," so to speak).

This much follows from a literal interpretation of the dependence thesis. But what would the father of the constitutive rule theory say about this? Searle is not an antirealist about institutional entities. In his writings he claims repeatedly and unequivocally that institutional entities are real, although they have a different mode of existence from natural entities: they are "ontologically subjective" or, in the terminology that I have used so far, mind-dependent. But dependent in what way, exactly? It is difficult to say, unfortunately, because in different parts of his writings Searle gives hints that point in opposite directions. Although many of his examples involve some form of causal dependence on human representation, elsewhere Searle claims that the constitutive rules theory must not be read in a robust psychological sense. The main point of the theory, he says, is to "lay bare" the "logical structure" of social reality. Institutional entities are dependent on representation regardless of what individual human beings actually think or believe about them. So it seems that Searle has in mind an ontological rather than a causal relation of dependence.

Is Searle willing to accept the odd implications of the thesis of noncausal dependence? Again it is not clear. Sometimes, especially when he is emphasizing the different modes of existence of natural and social facts, Searle appears to endorse infallibilism wholeheartedly. In a memorable paragraph, for example, he claims that we cannot be mistaken about the nature of a social event (a particularly lively cocktail party) once we have taken a certain collective attitude toward it. The social gathering is a party because we take it to be a party, *contra* realism as it is standardly understood:

> Something can be a mountain even if no one believes it is a mountain; something can be a molecule even if no one thinks anything at all about it. But for social facts, the attitude that we take toward the phenomenon is partly constitutive of the phenomenon. If, for example, we give a big cocktail party, and invite everyone in Paris, and if things get out of hand, and it turns out that the casualty rate is greater than the Battle of Austerlitz—all the same, it is not a war; it is just one amazing cocktail party. Part of being a cocktail party is being thought to be a cocktail party; part of being a war is being thought to be a war. This is a remarkable feature of social facts; it has no analogue among physical facts. (1995: 33–34)

One can hardly find a more clear pronouncement in favor of the dependence thesis. And yet, elsewhere in the same book Searle makes claims that go exactly in the opposite direction: the existence of institutional reality is not dependent on people's mental states.

> The process of the creation of institutional facts may proceed without the participants being conscious that it is happening according to this form. . . . In the very evolution of the institution [of, say, money] the participants need not be consciously aware of the form of the collective intentionality by which they are imposing functions on objects. In the course of consciously buying, selling, exchanging, etc., they may simply evolve institutional facts. Furthermore, in extreme cases they may accept the imposition of function only because of some related theory, which may not even be true. They may believe that it is money only if it is "backed by gold" or that it is marriage only if it is sanctified by God or that so and so is

> the king only because he is divinely authorized. (Searle 1995: 47)

So it seems that the members of the relevant social group can be mistaken, after all. Thinking that X is Y is not sufficient to make it Y. Or, going back to the previous example, the fact that the participants think of the Parisian bash as a cocktail party seems to be irrelevant, when Searle is in a realist mood. So it would seem that infallibilists and antirealists like Thomasson are pushing the constitutive rule theory beyond its intended boundaries.

Exegetical issues, however, are not our main concern here. The substantial question is whether the dependence thesis holds for social institutions or not. The thesis holds if it accounts for paradigmatic cases, and it fails if there are paradigmatic examples of institutional kinds that do not depend on anyone knowing the correct theory of the kind in question. It fails if some institutional kind exists in spite of generalized ignorance about the true conditions C that make X an instance of K. Such a kind would not be dependent on collective belief in a constitutive rule such as the one posited by Searle.

Although there are plenty of cases of this sort, for the sake of the argument it is wise to stick to a paradigmatic one, so we will have a deeper look at the conditions that ought to be satisfied for something to be money. As we have seen in chapter 3, we accept worthless paper bills in exchange for valuable goods or services because we are confident that we will be able to use them later to purchase other goods. The people who take my bills accept them as payment because they believe that others will take them, and so forth. Of course in principle different entities could fulfill the function of medium of exchange; but the process works a lot better if we all accept the same things, and we all share the same beliefs concerning what will be accepted in the future.

Such beliefs however should better not hang up in the air.

A primary role of the central bank is to ensure coordination among traders by enforcing a monopoly on the issuing of money. The bank prints bills that will be used by everyone, because everyone believes that the others believe—and so forth—that they will continue to be used as media of exchange. If an entity X (a paper bill) fulfills the condition C (being issued by the central bank) then it counts as money. But this means only that being issued by the central bank makes us very confident that the bill will be accepted in the future.

The whole thing looks suspiciously magical. Why should C matter? What is so special with a piece of paper that carries the stamp of the central bank? As we have seen in chapter 3, the conditions C have a facilitating role. But to function properly, a medium of exchange must satisfy other conditions that are not included in C. The key one as we saw is that everyone must believe that others will want to hold the paper bills in the future. But people must have good reasons (incentives) to hold currency. The state can guarantee a certain level of demand for the currency in the future, via taxation. If the state will only accept paper bills issued by the central bank as payment, then we can be confident that in the future people will have to hold at least some official currency for tax purposes. This is true of course only to the extent that the state is strong, is stable, and will have the means to collect tax. So, unsurprisingly, the strength of a currency depends on the political strength of the state.

The state collects paper bills via taxation, and puts them back in the economy by paying salaries to the employees of the public sector. If they meddle with the latter part of the cycle, though, governments and central banks can devalue the currency, creating inflation. In extreme cases of hyperinflation a currency may even become worthless paper. This is because to function properly a currency must be a reliable store of value. Standard economics textbooks remind us that the store of value condition is a fundamental presupposition for a currency to work as a medium of exchange. And this will be true only if the quantity of currency is relatively stable. So it seems that fulfilling conditions C is neither necessary nor sufficient. A cigarette can be money, even though clearly it has not been issued by the central bank, and a bill that has been issued by the central bank may fail to work as medium of exchange, if the state loses its credibility.

We now have the conceptual tools to debunk the dependence thesis. Notice that the store of value function—and other properties that back it up, like the stability of the quantity of money—is not included in C. But then being issued by the central bank is not what makes a token bill a member of the kind "money." The conditions C merely coordinate and support people's beliefs that individual paper bills fulfilling certain conditions will be accepted as means of payment in the future. And even this coordination function can be performed only under certain happy conditions. Money is constituted by a system of actions and beliefs in equilibrium, not by arbitrary conventions concerning the issuing of paper bills.

Another way to put it is to draw a distinction between currency and money. A currency, broadly understood, is a particular device that people use in a particular place and time to facilitate their trading activities. Dollars, drachmas, and sestertii are all currencies in this sense, but so are cigarettes and cowry shells in different societies and different times. The reason to distinguish currencies from money is that a given currency may or may not function as money, depending on whether it is able to do the things that money does (to work as a store of value, medium of exchange, unit of accounting). Although sestertii used to be money in Roman times, for instance, they are now just archaeological exhibits because they do not fulfill these functions anymore.

Of course people may still collectively decide to call sestertii "money" even though the equilibrium that allowed them to function as money in Roman times has collapsed. Perhaps when people see a coin, they say "it's money," even though they do not actually use it for trade. Suppose that they prefer to use shells as a medium of exchange instead. When asked what that thing they hold in their pockets is, they say "this is a shell," not "this is money." But functionally speaking, the shells are money and the coin is just copper.

The point is that one thing is to be recognized as money in a system of folk classification; quite another is to *be* money. The former does not imply the latter, contrary to what the constitutive rule theory suggests. Folk classificatory practices are in principle quite irrelevant. What matters is not what type of beliefs people have about a certain class of entities (the conditions they think the entities ought to satisfy to belong to that class), but what they do with them in the course of social interaction.

Counterexamples of this sort go against the grain of some of our intuitions, so let me elaborate a bit more. One reason why people are tempted to resist the conclusion of the argument, and maintain that a devalued currency is money, is that coins and bills are also artifacts. It is well known that artifacts tend to be identified on the basis of their intended function, that is, the function that was originally assigned by the producer of the artifact. This tends to happen regardless of the capacity of the artifact to actually fulfill the intended function. Thus a malfunctioning screwdriver is still a screwdriver according to our folk system of classification.

Now, a devalued currency is a malfunctioning artifact *and* a malfunctioning institution. Our intuitive system of artifact classification, therefore, prompts us to continue to classify it as money. But this trades on the confusion between currency and money that I have mentioned earlier. Although a currency is an artifact, being a currency is not enough for being money. The relevance of this distinction becomes apparent when we consider those rare cases of currencies that are not artifacts. Consider cowry shells for example: since they were not crafted with the intention of being used as money, our intuitive system of classification does not misfire. When they do not fulfill the functions of money, cowry shells are just shells.

We can use the same reasoning to see why other infallibilist examples are misguided. A party for instance is also a social artifact, because it is intentionally produced by someone's activity (the activity of the host). But what the host intends to achieve need not coincide with its final result. The reason why Searle's Parisian bash is not a party is that it does not fulfill the typical functions of a party. If I organize a social gathering to promote social bonding and entertain guests, but I do it in such a way that in the end everybody is bound to be miserable, then what I have organized was not a party after all. From an institutional point of view what really matters is not what we intend to do, how the gathering is called, or whether we believe that it is a party. What matters is what the gathering *does* for us (its real, rather than its intended function).

The decoupling of functions from people's intentions follows from the fact that institutions are systems of regulative rules. And for any particular set of rules, there is no guarantee that the rules are able to attain their intended goals. Suppose that, with the genuine intention to promote fun, we bring a lie detector to the Parisian bash. And suppose that, to have even more fun, we make it compulsory that the guests ask each other embarrassing questions concerning their extramarital affairs. If at the end of the evening "the casualty rate is greater than the Battle of Austerlitz" then clearly the format (the rules) that we have enforced have failed to attain their intended goal. How we decide to call the social gathering then is quite irrelevant: what matters is that the Parisian bash turned out to be an organized carnage, rather than a cocktail party.

The constitutive rules approach seems to imply that accepting X as Y immediately and necessarily makes it the case that the function Y is fulfilled, only because the regulative rules are left implicit in the standard formulation of the theory. But this is illusory, because the attribution of rights, duties, and the decision to follow a set of behavioral rules cannot by itself guarantee that any institutional goals are going to be attained.

So the real content of terms like "money" is not in the C conditions ("issued by the Central Bank"); it is in the actions ("accept it as payment") that are associated with the theoretical term ("money"). The kind money ultimately is nothing but this set of actions, and the related set of expectations. The C conditions are useful insofar as they simplify our decisions: they are coordination devices that help identify quickly and without lengthy inspection an appropriate set of actions in the given circumstances. (Should I accept a piece of paper as payment? Yes, because it has been issued by the central bank.) But to focus on the C conditions as what makes something a member of K is a mistake. It mistakes the coordination device for the system of actions and expectations that a social institution is.

Cases like this can be dealt with in three different ways. First, (1) one may deny that money is an institutional kind at all. Since infallibilism applies only to institutional kinds, it would be unaffected by cases like this. Alternatively, (2) one could accept that money is an institutional kind, but could save the dependence thesis arguing that collective acceptance must be interpreted in a weaker noncognitivist form. Or, finally, (3) one could reject the dependence thesis altogether.

Strategy 1 is a nonstarter, given that money is a paradigmatic social institution. The second strategy is to retreat to a weaker noncognitivist interpretation of collective acceptance: even though people do not explicitly accept the "right" conditions for K-hood, one could say that they do so *in practice*. Thomasson for example notices that social life often consists of a series of practices rather than conscious decisions based on explicit theoretical deliberation. Accordingly she suggests that the infallibilist thesis should be translated in a noncognitivist mode:

> Some might argue that in fact we seldom have explicit cognitive awareness of the relevant principles for institutional kind membership, we just have the practice of accepting certain sorts of things and rejecting others as

> putative kind members. I have spoken of the acceptance of principles in order to make the logical relations clearer, but the basic points can be made in a less explicit cognitivist scheme. The result in that case would be that even if (on a realist view) certain kinds of massive error in treating entities as members of a certain kind are possible for natural kinds (e.g., treating whales as fish), the same is not true for practices involving institutional kinds (e.g., treating cowry shells as money). (2003: 590n12)

Thomasson is saying that people may well be ignorant about the conditions that truly make X an entity of type K, and yet treat X exactly as a K-type entity. Infallibilism is turned into a thesis about *what we do* rather than about our explicit cognitive states: when the infallibilist says that we cannot fail to know that X is K, she really means "knowing how" rather than "knowing that." "Collective acceptance" becomes a theoretical concept, a state that is imputed by a theorist who observes a certain regularity of behavior. (Paraphrasing Dennett, we may say that the theorist takes the "collective intentional stance" toward a certain community and the behavior of its members.) But this does not imply that the actors must represent explicitly the constitutive rules that define the true conditions of kindhood.

This is a familiar move. Earlier in the book I have noticed that the concept of regulative rule can be interpreted in two different ways—as a representation that *guides* the behavior of the agents, or as a representation that is used by an observer to *summarize* and predict agents' behavior. The same applies here to the concept of constitutive rule: on the cognitivist interpretation, it is an "agent theory" that the actors use to guide and interpret their behavior; on the noncognitivist interpretation it is an "observer theory" that we use to make sense of what they do.

Is a switch to noncognitivism legitimate, or does it take all the interesting content away from infallibilism? Infallibilism now boils down to the claim that the members of a community are competent users of social objects. This effectively collapses the semantic and the epistemic challenges to realism: there is a boundary to K because de facto people use some entities, but not others, as K—even though they do not know why. By analogy with linguistic competence, if you speak a language correctly, then of course you know (tacitly) what is to be named as K and what is not—where "correct" is just what people in the community happen to call K.

But this noncognitivist version of infallibilism does not have any particular implications for the social sciences. The social sciences pursue knowledge in the sense of "knowing that," rather than "knowing how." And even from the viewpoint of the actors, noncognitivism drastically reduces the significance of the Error principle. From the fact that X is K we can infer only that people know *how* to sort the particulars that are K from those that are not; we cannot infer that they know *that* X is K.

Which leaves us with strategy 3. If you interpret acceptance in a strong noncognitivist fashion, then you must also refrain from deriving any substantial epistemic claims from the constitutive rule theory. This in turn implies that the odd couple should be abandoned. The best way to account for the existence of mistaken beliefs about institutions is to recognize that there are institutional facts about which we can be massively wrong.

This brings to completion the argument against dependence. The argument, to summarize, involves a number of steps and supports various conclusions. I started from the idea—very popular among philosophers and social scientists alike—that dependence on shared representations is a feature that demarcates social from natural kinds. In chapter 10 I have argued that the demarcation thesis can be made to work for at least a subset of social kinds that are involved in reflexive loops. However, I have also shown that the dependence thesis does not have any deep philosophical implications for the ontology and epistemology of social science. In particular, it cannot be used to question realism about social kinds. To reach this conclusion, I had to distinguish between two different interpretations of the dependence thesis. When dependence is interpreted causally, it can be shown that in many circumstances reflexive representations stabilize behavior. So dependence on representation may actually work in favor of real kinds. When dependence is interpreted noncausally (as ontological dependence) it does indeed lead naturally to endorse antirealism about social kinds. But as I have argued in this chapter, the idea that social kinds are ontologically dependent on (correct) representations is false. There are plenty of examples of paradigmatic institutions that exist quite independently of anyone holding any correct or incorrect theory about their functioning.

If the arguments that I have outlined are sound, then it follows that the differences between institutional and natural kinds are less important than many philosophers have surmised. The characteristic properties of institutional kinds ought to be discovered, just as in the natural realm. Membership in an institutional kind, moreover, is not a purely conventional matter. Conventions do play a role, but only in the choice of the devices that coordinate actions and beliefs. The choice of these devices may be arbitrary, within certain limits, but is not essential for the constitution of institutional kinds. The truly important properties—those that turn a token piece of paper into money, for example—are not conventional at all: they are functional properties, and they involve facts like the likelihood that others will accept paper bills in exchange for goods and services. These facts in turn depend on other facts like the number of bills circulating in the economy, economic growth, or the military power of the state. Collective attitudes toward the kind itself are neither necessary nor sufficient for the constitution of money.

There is a weaker, trivial sense in which conventionalism holds at a purely linguistic level. We cannot be wrong in *calling* X money, if X fulfills all the conditions that we accept as sufficient for the sort of things that the folk call "money." Our choice of language, or how we classify things for the purposes of everyday dealings and communications, is indeed a conventional matter. But what money *is*—the nature of K and membership in K—is not conventional at all. Satisfying whatever conditions people take to be sufficient for moneyhood does not necessarily turn X into an instance of money, unless other substantial conditions are satisfied. And thus, conversely, many things that do not satisfy the collectively accepted conditions may indeed be money.

Let us bid farewell to the dependence thesis, then, without any regrets, because what counts as an institutional kind is independent of our attitudes toward the kind itself.

REFERENCES AND FURTHER READINGS

Much of this chapter is based on Guala (2010, 2014). Searle's methodological remarks are scattered in his writings; on social ontology as "laying bare" the "logical structure" of reality, see, for example, Searle (1995: 90). Epstein (2015) argues that (what I call) the dependence relation is really the combination of two different metaphysical relations, which he calls "grounding" and "anchoring." On money, see the references at the end of chapter 9. Roversi, Borghi, and Tummolini (2013) have published the only empirical study of the conceptualization of social artifacts that I am aware of. There is some literature on malfunctioning artifacts in metaphysics and malfunctioning organisms in biology (see, e.g., Vermaas and Houkens 2003), but the problem of malfunctioning institutions has been largely ignored so far. The latter is reminiscent of a similar issue that arises in the philosophy of mind (the so-called problem of "mad pain"; see Lewis 1980), although the solution that I propose here differs from the canonical one (I am indebted to Mikael Cozic for pointing this out).

CHAPTER 13

MEANING

In the previous chapters I have defended realism about social kinds: the properties of social kinds or institutions must be discovered a posteriori, by means of empirical investigation. It is not something that we can know merely by participating in a social practice, as some philosophers have claimed. I have also argued that realism comes at a cost: one of its implications is that we can be wrong about social kinds. One may be a competent speaker, and apply correctly a term like "money," without actually knowing what money is. Realism and fallibilism go hand in hand because they are two sides of the same coin.

In this chapter I explore some implications of realism, looking at cases in which our understanding of institutions has nontrivial consequences for the way in which we live and organize our societies. I believe that in some circumstances ontology can give significant contributions to the progress of a society, bringing clarity in debates where positions are polarized and language is charged with controversial connotations. In particular, I will examine current controversies on same-sex marriage and show that the theory outlined in this book may help us see the problem from a new perspective.

Contemporary debates about same-sex marriage are a curious mix of politics, ontology, and semantics. The issue is obviously political, because it concerns the extension of certain rights (such as the rights to inherit property and to have children) to gay and lesbian couples. But it is also ontological because, according to some philosophers, politicians, and religious leaders, a marriage cannot by its very nature be a union between two individuals of the same sex. And finally, part of the controversy is semantic, for it concerns the correct use of the term "marriage."

Since in the course of the chapter we will have to delve into philosophically complex issues of meaning, it is useful to start with a brief introduction to semantics. Contemporary versions of realism are tightly related with *externalism* about meaning. The core of externalism is usually summarized in the slogan that "meanings are not in the head" of the speakers. The slogan can be cashed out in different ways, but a brief sketch will be enough for our purposes. Externalists distinguish between the *extension* of linguistic terms and the *theories* or beliefs that people hold about the putative extension of the terms. For example, many people think of water as a transparent, colorless, tasteless liquid that we can drink to quench our thirst. This is the *stereotype*, the "conception," or the "folk theory" that people associate with the term "water." The stereotype is the part of meaning that most people acquire when they become competent speakers. But the folk conception by itself cannot determine what the term "water" refers to, because people's knowledge of the nature of water is fallible and imperfect. In fact, the stereotype can be (and often is) revised in light of new discoveries that change our conception of the thing that the term is used to label.

According to realists and externalists, what water is depends on the way the world is. Over time people delegate to experts (scientists) the study of the stuff that they call "water," hoping to find the properties that determine its identity as a kind. The experts typically formulate a *scientific theory* that departs somewhat from the stereotype. Like the stereotype, the scientific theory does not determine what "water" means, but provides the best approximation that is available at a given time, and that is used to adjudicate controversial cases (what is water and what is not) in specific circumstances. So, for example, water refers, to the best of our knowledge, to a substance constituted by two atoms of hydrogen and one of oxygen. But again, the extension of the term is not determined by this theory, which may turn out to be wrong after all; it is determined by the way the world is, by all the things that share the properties that constitute the substance that we call "water."

To cut it short, the process of meaning attribution consists basically of three steps: First, a sample of paradigmatic objects or substances are identified on the basis of some similarity criterion and are labeled with a term (baptism). Then, experts engage in a systematic search for the relation of "sameness," or in other words they try to identify the properties that are shared by the paradigmatic objects (discovery). Finally, the reference of the term is refined to include those entities and only those entities that have these properties (application).

In the case of social kinds, it often happens that the scientific theory differs significantly from the folk theory. People have all sorts of extravagant ideas about what makes Elizabeth Mary Windsor the queen of England, or what money is, or what is marriage. These ideas often turn out to be false after careful investigation, when it turns out for example that what people actually do is different from what they think they do (the stereotype). When the social practice is inconsistent with the stereotype, the stereotype must be revised or abandoned in favor of a more adequate theory. However, it is quite common for folk theories to survive alongside scientific theories, because they perform well enough as guides for the everyday application of social terms. So it is not rare for the same institutional term to live a double life, as a theoretical term associated with a scientific conception embedded in a scientific theory, and as an everyday term associated with an informal concept embedded in a folk theory.

In a series of essays, Sally Haslanger (2012) has introduced a useful distinction between *manifest* and *operative* concepts.

The operative concept associated with a given term is the concept that is implicit in our practices. For example, if we use cigarettes primarily as a medium of exchange, then the operative concept of "cigarette" is, effectively, money. The manifest concept in contrast is the concept that is consciously or explicitly endorsed by most people. So, for example, we may think of cigarettes primarily as rolled tobacco to be smoked. According to externalism, the identity of an institution and the meaning of institutional terms is determined by the operative, not by the manifest concept. If we want to know what an institution is, we must study people's practices, not their folk theories. So what cigarettes are in a certain community (money) is not determined by what people think they are (rolled tobacco for smoking) but by what they do with them (they use it as a medium of exchange).

The scientific study of practices clearly is a major source of information about social reality. But is knowledge of the practice *sufficient* to determine what an institution is? To say so would have rather controversial implications. Consider for example the case of marriage. Most people in our culture conceive of marriage as a formal union between an adult male and an adult female. And a scientist studying contemporary marriage practices in Western countries would conclude that they primarily regulate relationships between adult males and females. So in this case the stereotype and the practice, the manifest and the operative concepts coincide. But if this is what marriage really is (a contractual relation between two individuals of different sex) then the term "marriage" should not be used to refer to those contracts that regulate unions between two women or two men. We should use a different term for different institutions, in this case for contracts among adults of the same sex.

This argument is not a mere philosophical quibble. On the contrary, it has great practical relevance, and has been used repeatedly in political and legal controversies. In many countries gay and lesbian activists are fighting a battle for the extension of marriage to same-sex couples. After much resistance, the activists are beginning to see the fruits of their campaigns: unions between individuals of the same sex have been legalized in several countries, including the United Kingdom, Canada, France, Brazil, the Netherlands, and Spain. In some of these countries, however, legislators have decided not to use the term "marriage" for these contracts. One reason is that same-sex unions (like the French *pacte civil de solidarité*, or PACS) are more acceptable to conservative voters and politicians if they are kept formally distinct from traditional marriage.

A prominent argument employed by conservatives is that we should use different names for different things; "marriage" should not be used to name something that differs in relevant respects from traditional marriage. New things deserve to have new names. In November 2009, for example, Archbishop Joseph Kurtz, chair of the Ad Hoc Committee for the Defense of Marriage, summarized the position of the Catholic Church as follows: "We work to preserve the true meaning of marriage. Especially in our society where we see so many marriages fail, we should work to strengthen marriage rather than redefine it. Marriage must be protected and promoted today for what it is and what it is meant to be: the lifelong, exclusive union between husband and wife" (United States Conference of Catholic Bishops 2009).

One can hardly be blamed for seeking clarity, and the principle "one name, one institution" seems reasonable enough in principle. The worry is that extending the notion of marriage to include same-sex couples, and allowing them to engage in practices that were traditionally reserved for heterosexual couples would cause a shift in the meaning of marriage. If institutions are rules, changing the rules would change the institutions. So new institutions require new names to avoid confusion.

Many gay and lesbian activists see the introduction of contracts like PACS just as a preliminary step. The use of a different terminology, they argue, reflects a prejudice against homosexuals and is effectively a form of discrimination against same-sex couples. For this reason, they have asked the judicial bodies that oversee the application of constitutional rights to extend the legal use of the term "marriage." In 2001 for example the Supreme Court of British Columbia was asked to examine the constitutional legitimacy of denying the use of the term "marriage" for legal unions between spouses of the same sex. The *Halpern v. Canada* appeal was similar to many trials that had taken place in other countries, with one significant difference: the Supreme Court availed itself of the opportunity to call several scholars as expert witnesses. The range of expertise ranged from history to anthropology and feminist studies, but semantic arguments featured prominently in the affidavits of two philosophers of language, Robert Stainton and Adèle Mercier.

Robert Stainton, a professor at the University of Western Ontario, argued against the extension of marriage to same-sex unions. The arguments that he submitted to the court were meant to support the conclusion that calling "marriage" the union of two men or two women would constitute a misapplication of the term:

> I believe that a correct understanding of the semantic content of the term "marriage" does not permit us to ask "whether or not same-sex couples should be allowed to be married." The correct semantic analysis suggests that we cannot sensibly ask whether men should be able to marry one another, any more than we can ask why two boys cannot be sisters, or why a bachelor cannot be married. The current semantics of our common language simply rule out these options. It is not a question of some official body permitting boys to be sisters, or not permitting bachelors to be married. Sisters are women, and bachelors are unmarried, as matter of present meaning. In the same way, I will suggest, matrimony just is the union of a man and a woman. (Stainton, para. 10, quoted in Mercier 2007)

Although the semantic argument against marriage predates *Halpern v. Canada* by decades, I will focus on Stainton's affidavit because it is a rare attempt to formulate the argument explicitly. Its basic structure is the following:

1. The term "marriage" means "the union of a man and woman."
2. Therefore, it cannot be used to name temporary, nonexclusive, same-sex unions.
3. To use the term otherwise would be to change the meaning of "marriage."
4. Since this would cause confusion, we should better call same-sex unions using a different term ("Civil Unions," for example).

Notice that the semantic argument properly construed applies to the term "marriage," rather than to the institution. But since the argument relies on the identification of the meaning of "marriage" (first premise), it can be made to work only if one is able to demonstrate that marriage is the union between a man and a woman. It requires that we know what marriage as an institution is.

How can we identify the nature of an institution though? Stainton endorses a view of institutions as practices. He argues that the meaning of "marriage" is determined by the practices that are traditionally or historically associated with the term "marriage." And these practices, he claims, are legally sanctioned unions between a man and a woman:

> The key distinguishing feature of marriage is its history, including in recent history, its religious origin. This history is reflected in the highly specific ceremonies by which couples marry. It is true that one can now undergo a civil ceremony, and that such ceremonies were historically available, but even current civil ceremonies are clearly a descendent of older, specifically religious traditions. . . .

> To look at marriage as divorced from the religious interpretation of the practice and the historical background of the practice is, in my opinion, to miss the meaning of marriage altogether. Marriage just is a descendent of a highly particular, highly stylized practice with religious roots. To divorce marriage from its history in this way would be like trying to divorce being a nun or being an altar boy from their religious roots! Though the institution has been embraced by the State and marriage has a legal status, this has not erased its origins. So, when I say that marriage "is" a descendent of its history, this is essential to understanding precisely what the word "marriage" actually means. We cannot divorce our understanding of the term from its history. (Stainton 2001, paras. 19–20; quoted in Mercier 2007)

Although it recurs frequently in semantic arguments, the claim that marriage has religious origins is at least controversial and most likely untrue. However, I will bracket this issue for the time being. The important point is that according to the opponents of same-sex marriage (for simplicity, I will refer to them as "traditionalists") the practice defines the institution and, a fortiori, the meaning of the institutional term. To understand what marriage is and what the term "marriage" means, we have to look at the way it has been used up until now.

And yet, many people believe that the term "marriage" can be legitimately applied to gay and lesbian unions, because the sexual identity of the spouses is not essential to marriage. Marriage for these "reformists" is a contractual relationship between two persons, independently of their gender. So who is right and who is wrong about this matter? Is it a purely verbal dispute, or is there something more at stake?

I will argue that the disagreement about the meaning of marriage is not purely verbal, and cannot be resolved by simply stipulating that marriage is this or that. For a realist, the meaning of marriage is determined by the way the world is, and the language that we use must be true to the way in which the (social) world is organized. So the problem for a realist and a reformist is how to justify the claim that marriage does not exclude gays and lesbians, in a world in which as a matter of fact marriage is overwhelmingly restricted to heterosexual couples.

The theory of institutions as rules in equilibrium, which I have outlined in the first part of the book, suggests that we should not give too much weight to the fact that a certain institutional arrangement has been prevalent in a society for a given period of time. If institutions are solutions to problems of coordination, then there should be other rules that solve the same problem in a different way. The historians and anthropologists who have described the functioning of alternative institutions, in fact, have always been an important source of inspiration for the critics of traditional customs.

One possible approach then is to challenge arguments like Stainton's empirically, by showing that there have been cases of same-sex marriage in our culture. The historian John Boswell (1994) for example has argued in a controversial book that same-sex marriage was practiced in antiquity and during the early Christian era. Even Roman emperors, apparently, did it: Nero for example is said to have married a male partner at least twice, in separate occasions. During the early Christian era, moreover, the Church recognized the institution of *adelphophilia*, a relationship between two adult males based on spiritual affinity, formally acknowledged during a public ceremony that (according to Boswell) was very similar to a wedding.

The empirical challenge, however, faces an obvious difficulty: it is possible to reply that same-sex unions are anomalous, and that they cannot be considered genuine instances of marriage. Several historians for example have pointed out that Roman law required different genders for *matrimonium*, and Boswell's central example (*adelphophilia* or "brotherhood") was kept formally distinct from *matrimonium*. Nero was an exception because, as emperor, he could do whatever he wanted. One of his predecessors, Caligula, had appointed his horse a priest, after all, and had promised to make him a consul. But most people would agree that a horse cannot possibly be a priest or a consul. By analogy, one may argue that Nero's eccentricities are not counterexamples to the thesis that marriage is a relation between a man and a woman.

I will refer to this objection as "Caligula's problem." Although I do not think that arguments based on Caligula's problem are conclusive, they highlight a difficulty that mars the empirical strategy. The difficulty consists in sorting genuine counterexamples from anomalies that should be treated as cases of "pseudo-marriage." An externalist, who holds that the meaning of a term is determined by the way in which the world is, should be committed to the use of empirical evidence for the identification of the true meaning of marriage. But Caligula's problem stands in the way of an empirical solution, and requires an adequate response. As we shall see later, empirical evidence is important, but it plays a different role in the identification of what marriage is or could be. Before we come to that, however, it is worth examining another approach that accords a limited role to empirical evidence and emphasizes the importance of normative considerations.

One reason why many reformists think that the traditionalist position is unacceptable is that gays and lesbians *should not* be excluded from the institution of marriage, whatever shape that institution might have had in history. This intuition in turn is backed up by the consideration that institutions are not static entities: they evolve, and there must be some way in which they can be modified to serve our changing needs. So the question is not primarily what our institutions are *now*, but what we want them to be or become in the future.

The case for the inclusion of normative considerations appears increasingly plausible if we focus on examples that are formally analogous to the case of same-sex marriage: imagine a social scientist studying the institution of parliament in 1900, for example. Having traveled around the world the scientist reports that most members of parliament are adult males. He also observes that being an adult male is used pretty much everywhere as a necessary condition for being elected in parliament. Therefore, he concludes that the institution that people call "parliament" is an intrinsically male institution. Or imagine a social scientist visiting the United States in 1900 to study the institution of marriage. She finds that white people never marry black people, that miscegenation is a taboo, and that there are laws that formally prohibit interracial marriage in most states. She concludes that the institution of marriage is restricted to members of the same race.

Most of us have the feeling that these scientists would be wrong: it is ludicrous to infer from these cases that marriage is *necessarily* restricted to members of the same race, or that parliaments are for males only. But why do we think that? Surely we are inclined to reject a narrow definition of these institutions for political reasons: we believe that these restrictions are wrong, and we do not care whether a certain (descriptive) theory of parliament coincides with current practices. Institutions can and sometimes ought to be designed differently so as to fit our normative desiderata.

This line of argument has been pursued by social constructivist scholars. The argument revolves around an ontological claim about the nature of marriage, namely that it is a social and dynamic entity without a fixed essence. This thesis is then used to derive the conclusion that there are no external limitations to the way in which we can use the term "marriage." As a prominent feminist scholar has put it, "there is no 'always has been and ever shall be' truth of marriage" (Hunter 1991: 17). The only limits are internal to the political community, because the definition of institutional terms reflects ultimately the power relations between different parties. William Eskridge, for example, characterizes the constructivist position by means of three claims or hypotheses:

> First, marriage is not a naturally generated institution with certain essential elements. . . . Second, the social construction of institutions like marriage is not and cannot be neutral, for it involves the playing out of a society's power relations. . . . Third, the social construction of marriage is dynamic. . . . Marriage should change as the subordinated groups identify their own oppression and decide to resist it. (1993: 1434)

Sally Haslanger similarly has argued that normative considerations play a legitimate and important role in the analysis of concepts like "race," "gender," or, indeed, "marriage." The project that she has variously called "analytical" or "ameliorative" requires

> to consider what work we want these concepts to do for us; why do we need them at all? The responsibility is ours to define them for our purposes. In doing so we will want to be responsive to some aspects of ordinary usage (and to aspects of both the connotation and extension of the terms). However, neither ordinary usage nor empirical investigation is overriding, for there is a stipulative element to the project: *this* is the phenomenon we need to be thinking about. Let the term in question refer to it. On this approach, the world by itself can't tell us what gender is, or what race is; it is up to us to decide what in the world, if anything, they are. (Haslanger 2012: 224)

An analysis of social institutions, according to Haslanger, requires that we pay attention to three elements: (1) the "operative concept," (2) the "manifest concept" or theory, and (3) the "normative concept." The operative concept is the practice or set of practices that is associated with the use of a term; the manifest concept is the theory or stereotype that people use to make sense of the term; and the normative concept is the target, that is, the institutional entity as we would like it to be. The three concepts often overlap to a large extent, but inconsistencies are not rare. When manifest and operative concepts are at odds, for example, an empirical study of the way in which people use a given term may lead us to revise the manifest concept that we associate with that term. We may discover, for example, that the term "homosexuality" does not refer primarily to a biological kind, but to a set of behaviors that are considered inappropriate for moral (rather than, say, medical) reasons. But it is also possible that while the operative and manifest concepts coincide, they both disagree with the normative target that we have set for ourselves. In such cases, we must ask a battery of questions at once:

> What policy do we want to promote . . . , and what do we want to do with the bit of language we have been using? Do we want to change our policy and keep the same term, change it and introduce a new term, keep the policy and change the term, or keep the policy with the old term? (Haslanger 2012: 379)

The goal is to achieve consistency among the three elements —a sort of reflective equilibrium, in Rawlsian terms. So we may be entitled to reject a descriptively correct theory if we dislike its normative implications. This way Haslanger tries to save the entrenched intuition that morally unacceptable institutions can be reformed, without giving up realism about social kinds. According to Haslanger "there is a central form of social construction . . . that is compatible with important forms of realism, an objectivism about kinds, and naturalism" (2012: 183). One can be a social constructionist, because gender and race are socially maintained, rather than biologically determined, categories; and one can be a realist, because the terms "race" and "gender" refer to real social kinds, that is, classes of entities that share important properties and whose commonalities are sustained by clearly identifiable social mechanisms. "Both gender and race are real, and both are social categories" (Haslanger 2012: 246).

Haslanger's position is interesting because there is an obvious tension between realism and constructivism. How can one simultaneously be a realist about gender and race, and hold that "the world by itself can't tell us what gender is, or what race is"? To be a realist is precisely to believe that reference is determined by the world, and it is not "up to us to decide what in the world, if anything, they are." According to the realist, we cannot change our institutions freely: the identity of any entity is robust to a limited range of manipulations only, so there are changes to an institution that will turn it necessarily into something else. Just as the term "senator" cannot be stretched so far as to include Caligula's horse, so there are arrangements that cannot be legitimately called "marriage," regardless of what we want the latter to be. The problem is to find what the boundaries of an institution are, and to identify the meaning of "marriage" accordingly.

Haslanger does not say much about the correct way to resolve this tension. What she does say, however, points us in the right direction. The normative element in the ameliorative approach invites to redefine institutions so that they better serve our goals and purposes. But the realist element reminds us that the goals and purposes an institution serves are constrained by properties and mechanisms that are not fully under our control, because they are determined by the external world. In the next chapter I will try to articulate these two insights and to provide a theoretical foundation to Haslanger's ameliorative project, using functions as identity criteria for a realist theory of institutions.

Before I do that, let me just add that taking realism seriously, as Haslanger encourages us to do, has various advantages. Radical constructivism—the idea that there are no external constraints on the definition of institutional concepts—creates problems that we would be hard-pressed to solve. For instance, suppose we wanted to reconceptualize marriage in such a way that it fits "our" normative requirements: whose normative requirements should be taken into account, exactly? Marriage is a contested concept not just or not primarily empirically, but also morally and politically. The traditionalists who oppose same-sex marriage typically use a mixture of factual, metaphysical, and normative arguments, where moral judgments play a large role. For instance, here is Archbishop Kurtz again:

> Protecting marriage affirms the unique and indispensable roles of mothers and fathers, and recognizes the particular responsibilities that husbands and wives bear in society. Protecting marriage affirms the permanent and exclusive love between a husband and a wife as a wonderful and incomparable good in itself which also is of great social and practical consequence. Their sexual difference, man to woman and woman to man, is real and valuable—not a social construct, and not an aspect of the human person that may not be disregarded at will and without cost. This difference is essential for marriage and is the relational context for the formation of the human person. Sadly, the attempts to redefine marriage today ignore or reject the unique identity and gifts of man and woman. Such a dismissal only fosters confusion about what it means to be human. (United States Conference of Catholic Bishops 2009)

The man-woman relation is not only "real" and "essential," but also "valuable," "a wonderful and incomparable good" that "may not be disregarded without cost." So traditionalists on the one hand and reformists on the other disagree first and foremost on the normative principles. The issue is whether the reflective equilibrium must be shaped on our normative intuitions or theirs, and there does not seem to be a lot of room for rational agreement on these matters.

So it would be a mistake to let normative considerations alone determine what an institution is. There are certain matters of fact regarding what an institution is or can be, and people with different normative convictions can have a rational conversation about such matters. A realist approach should allow one to determine whether marriage does or does not exclude gays and lesbians as a matter of fact, not just as a matter of moral conviction. It is not a question of tradition, or people's understanding of marriage: the issue is whether heterosexuality is or is not part of the nature of marriage. In the rest of the book I will explore a solution to this conundrum that saves both the doctrine of realism and the normative intuitions that most reformists have about the nonexclusive, nondiscriminatory nature of marriage.

REFERENCES AND FURTHER READINGS

Eskridge (1993) reviews briefly the use of semantic arguments about marriage in the legal literature. In the political arena, the argument featured prominently in the campaign for the defense of marriage led by President George W. Bush in the first decade of this century (see, e.g., Duncan 2009). Putnam (1975), Kripke (1972), Brandom (1994), Horwich (1998) defend different externalist accounts of meaning, while Devitt and Sterelny (1999) offer a clear synthesis of the philosophical debate. Although I use the standard example of "water" in the main text, there are reasons to doubt that water is actually H_2O (see Weisberg 2006). Boswell's (1994) controversial book generated a huge secondary literature, but Rapp (1997) is a good place to start. The analogy between gender and race discrimination has been widely used by legal scholars; see, for example, Law (1988) and Koppelman (1988). On reflective equilibrium as a mode of justification, see Rawls (1971) and Daniels (1996). The *Halpern v. Canada* case is an excellent example of how philosophy can occasionally enlighten intricate and controversial social issues.

Some of the legal material is available at http://www.sgmlaw.com/en/about/Halpernv.CanadaAttorneyGeneral.cfm. The affidavit of Robert Stainton is not available online, so I have quoted from secondary sources. The affidavit of Adèle Mercier can be found at http://www.academia.edu/Documents/in/Adele_Mercier. Her philosophical arguments are presented and articulated in more detail for an academic audience in Mercier (2007).

CHAPTER 14
REFORM

The previous chapter left some questions unanswered: Is it possible to be simultaneously a realist and a reformist? Is it possible to believe (1) that the nature of social institutions is determined by the way the world is and (2) that the institutions that regulate our lives can be changed without turning them into different things? In this chapter I argue that it is possible, and propose a solution that saves at the same time the externalist commitment to let the world determine the meaning of institutional terms, and the intuition that institutions can be modified to fit our normative desiderata without turning them into something else.

The solution that I propose is an improvement with respect to the "ameliorative" approach pursued by Haslanger (2012). The ameliorative theorist believes that she must challenge conservatives on normative grounds because there is no room to attack them descriptively. Take the case of marriage again: if theory and practice did not coincide, we could highlight the inconsistency and ask the conservative to be faithful to their own implicit understanding of marriage. But what if "marriage" has always referred to same-sex unions in a given culture? When the empirical strategy does not work, the ameliorative approach offers an attractive solution. We ought to criticize current practices normatively because there is no way to be a reformist within a purely descriptivist framework.

In this chapter I take a different route. My view is that the consistency between theory and practice (or operative and manifest concepts, in Haslanger's terms) is not a hallmark of correctness: a theory may be descriptively false even if it is consistent with people's practices. The underlying reason is that current and past practices are not all the evidence that we can have to assess our theories. As philosophers of science know very well, theories are *underdetermined* by existing data: no matter how close a fit between the theory and the existing evidence, it is always possible that the theory is unable to account for some data that have not been considered yet. So the fact that marriage has been applied predominantly to male-female relationships until now does not imply that marriage is limited to such relationships. It may just be that we have not collected enough evidence, or evidence of the right kind; if we had done it, we could have discovered that the institution of marriage encompasses same-sex unions after all.

Recall the externalist's story about the discovery of meaning: the revision of a stereotype begins with the identification of a sample that scientists study to identify the characteristic properties of a real kind. But not any sample will do: if the sample is unrepresentative of all the things that belong to that kind, we may fall into error. We may wrongly infer for example that "all swans are white," because we have not taken Australian swans into account. So the problem with the theory "all marriages are heterosexual" is that it may be based on a narrow evidential basis. This is why a standard strategy of gay activists has been to enlarge the sample and search, à la Boswell, for cases of same-sex marriage through history and across cultures.

But beware: the claim that a sample is inadequate cannot rely *only* on evidence, for the reasons that I have mentioned in the previous chapter. Faced with the objection that same-sex marriages were common in, say, ancient Rome, the traditionalist may reply that they were not genuine cases of marriage—they were something else. This would be analogous to say that black swans are not real swans. But in the case of swans we all agree that it would be an unacceptable ad hoc move. Why? Because there are good *theoretical* reasons to classify white and black swans as two species (*Cygnus cygnus* and *Cygnus atratus*) within the same genus (*Cygnus*). Both species descend from common ancestors, for example, and share a cluster of traits that have been shaped by natural selection.

So we need a theoretical reason to reject similar moves in the case of marriage. Here scientifically-informed ontology comes to the rescue. The inference from practice (all marriages are heterosexual) to theory (marriage is heterosexual) trades on a confusion between types and tokens. The institution of marriage in the West, or in any historically existing culture for that matter, is not marriage in general. It is a particular instantiation (token) of the type of institution that we call "marriage." Recall that according to the unified theory an institution-token is a particular solution to a set of related problems of coordination. So by studying marriage practices in Florence during the thirteenth century, say, we can only learn about the particular way in which Florentine people organized child-rearing, reproduction, and economic cooperation at a particular time. To find out what marriage is *in general*—what marriage is *as a type*—we need to focus on the coordination problem as a whole, rather than on its specific solutions, because institution types are defined functionally, by reference to their goal or to the problems they solve. Since the same goal can be attained in many ways, the study of specific institutions tells us more about the specific solutions than about what they have in common. What all the institutions share is that they are solutions to the same problems, or equilibria of the same class of games.

Describing Florentine marriage as a token institution may seem counterintuitive, so it is worth spending a few words of explanation. Whether an entity or set of entities is classified as a type or a token depends on whether it is located in space and time. Tokens are concretely instantiated in some geographical location and have a history, a beginning and an end. To treat Florentine marriage as a token is consistent with this approach, because that specific contractual solution is a historical instantiation of the general institution of marriage (the type). This is analogous to the way in which many biologists and philosophers treat species: like organisms, species are particulars located in space and time. Higher taxa, in contrast, are types or universals. An important motivation behind this distinction is that types are not involved in causal processes in the same way as individuals do. They do not interact with each other, for example, or with other entities. So tokens seem to enjoy a kind of existence that types lack. While marriage as a type is causally inert, marriage according to Italian law affects the behavior of Italians and is (arguably) affected by it.

Whether the analogy holds perfectly, of course, may be disputed. However, nothing hinges on this particular choice of classification: the main point is that social entities are stratified at different ontological levels, and that we should better keep these levels separate. The problem, as far as institutions are concerned, is that social scientists have not developed a taxonomy, and hence a terminology, that allows to make these distinctions clearly. By means of "institution-type" and "institution-token" I am trying to distinguish different levels of ontological analysis using the limited linguistic resources of contemporary philosophy and social science.

The important point, in any case, is that we cannot determine what marriage is (the type) simply by generalizing across a sample of token institutions, because the sample may be inadequate. If for contingent reasons people have always converged on the same solution, it does not follow that no other equilibria are possible. As an analogy, consider a chemical element like uranium-239. For contingent reasons, such an isotope did not exist in nature before scientists learned how to produce it in a nuclear reaction. And we can easily imagine an alternative scenario where humans have never learned how to do it and uranium-239 has remained a mere possibility. But in this scenario clearly it would have been false to infer from the fact that such an isotope had never been observed, to the theory that uranium has isotopes ranging from 217 to 238 only. Just as we need theoretical physics to tell us which isotopes *can* exist, and how to classify them, so we need a theory of institutions to tell us which marriages can exist, and how to classify them.

Another way to put it is that token institutions (marriage in ancient Rome, Muslim marriage, Tibetan marriage, etc.) may be used as defeasible evidence to develop an adequate theory of the institution type (marriage in general). But the theory does not have to fit the practices that are instantiated in token institutions narrowly. By observing different solutions to the same problems of coordination we are alerted to the fact that we are dealing with a game that has multiple equilibria. This claim must then be substantiated by other means, for example showing that the problems that people solve using a certain contractual form are similar in some respects, that the incentives are structured in a certain way, and so forth. Finally, it takes theoretical ingenuity and imagination to discover some nonobvious equilibria that may theoretically exist but have never been implemented before.

In the case of marriage, the various rules that people have used to regulate their relationships typically solve several problems at once. Some are problems of procreation—how to regulate the production of children. Others are problems of education: how to raise, feed, teach, socialize children until they become adults who are well adjusted in their community. But there are also issues of economic cooperation (who should help whom and how, who should inherit what), mutual emotional and affective support, and so on and so forth. Historical and anthropological evidence shows that none of these problems are essential, although they constitute a "cluster" in the Boyd-Mill sense. There are couples who raise kids they have not biologically conceived (adoptive and foster parents); couples who help each other economically without raising children; couples who raise children and share economic resources without giving each other emotional or affective support. Marriage does not have a single essential function. But as long as a contract regulates (formally or informally) a cluster of activities of this sort, it is legitimate to speak of marriage.

Contrary to the ameliorative approach, this strategy has the advantage of saving a key realist principle: what marriage is, or what deserves to be called "marriage," does not depend on our subjective beliefs or intentions. It does not depend on what people want, or on their normative desiderata either. It depends on how the world is, in particular what kind of problems the institution of marriage (in its various instantiations) solves. This is consistent with the spirit of meaning externalism, and opens the space for a descriptive, scientific approach to the issue of same-sex marriage. There is a big difference between saying "this is what we want marriage to be" and "this is what marriage is." In the former case, the ontological conclusion is a direct consequence of political negotiation, or the balance of power; in the latter, it is a consequence of empirical and theoretical investigations in the institutions that people use, have used, and can use to solve their problems. It saves the realist principle that what marriage is, is primarily a scientific issue, and helps to keep the latter distinct from the political question of which one, among its many possible versions, we want to adopt.

Thus normative considerations do and must play a role in political debates about institutional reform. The point is that normative considerations play a legitimate role only when token institutions are concerned. The question of what marriage is in general must be kept distinct from the question of which version of marriage we ought to adopt, because each answer must be justified in a different way, using different methods of investigation, and relying on a different kind of expertise.

If K is a natural kind, the experts to whom we must delegate the task of investigating and classifying natural entities are natural scientists (biologists or psychiatrists, for example). If it is a social kind, in contrast, the experts are anthropologists, sociologists, or economists. Finally, if K is a normative kind (a legal, religious, or moral kind, for example) we may want to delegate classificatory decisions to judges, priests, and moral philosophers—if we trust their expertise. Or we may let the citizens decide, as it is customarily done in modern liberal democracies. But in any case it is important that we have very clearly in mind what kind of question we are asking, and who has the competence to answer it.

Being able to distinguish between different kinds of kinds is crucial to understand the controversies that erupt around the use of contested terms like "marriage." Recall the case submitted to the Canadian Supreme Court: if marriage is primarily a social institution, then the experts on the nature of marriage are social scientists. They have the expertise to decide (on the basis of scientific evidence and theory) what marriage is as a social institution. But they do not have the right to decide what marriage is or should be from a legal point of view—that is, how to codify marriage in the law. This is a normative juridical matter, and social scientists are not experts in this regard.

Social scientists can help make distinctions between different uses of the term "marriage," for example when we use it to refer to the institution in general, versus some of its specific (token) instantiations. And they can clarify that the institution of marriage in general is not limited to heterosexual relations. They may even suggest new ways to regulate marriage relations that have never been experimented before, by way of theoretical speculation. But they cannot (qua scientists) tell us which of these solutions ought to be incorporated in our legislation.

Adèle Mercier (2007), who participated as an expert witness in the *Halpern v. Canada* case about same-sex marriage, has introduced the category of *legal kind* to mark this distinction. Decisions about the legal notion of civil marriage may be delegated to the citizens, for example through a referendum or through the decisions of their representatives in parliament. If this route is followed, the issue will be decided by majority vote, and the outcome will reflect the majority of people's convictions (or the convictions of their delegates) about the way in which marriage ought to be regulated in their country. But this would not be a judgment about the nature of marriage qua social kind, because that's something about which the members of parliament and even the citizens themselves do not have the authority to legislate. So even if same-sex unions were prohibited by law in Canada, it would still be legitimate to refer to them as "marriages," as a possible alternative institutional arrangement to the one that has been adopted in the legislation.

Another option is to delegate the decision to a body of legal experts, such as the Supreme Court. This is the procedure that has been followed in Canada and in many democratic countries that care to protect the rights of individual citizens from discrimination or the oppression of the majority. In British Columbia the Court was called to decide whether denying usage of the term "marriage" for same-sex contractual relations would constitute discrimination against gays and lesbians. The traditionalists' case was in part supported by the arguments that I have reviewed in the previous chapter: plaintiffs claimed that the term "marriage" refers to a practice that is historically limited to heterosexual couples, and argued that to make it more inclusive would change the meaning of the term. The Court eventually rejected these arguments, and deliberated that the use of the term "marriage" for same-sex contractual relations is not only legitimate, but mandatory to avoid discrimination.

Why did they do it, using what methodology, and on the basis of which expertise? Mercier points out that the Court did not deliberate on whether same-sex unions should be regulated or not by means of a legally recognized contractual relation. And it did not deliberate on what marriage is qua social relation. It just determined that to restrict the term "marriage" to heterosexual unions would discriminate against gays and lesbians. Or, more precisely, that the identification of legal marriage with a specific concept of marriage (marriage for heterosexuals) would clash with the freedoms and rights recognized in the Canadian constitution.

> The Court has no mandate to change or to prescribe the ordinary meanings of words. Its mandate is to individuate them, and to identify which may figure in our Laws. It is the Constitution, in the last instance, which determines the same-as relation for (civil) marriage. The reason for this is because, to the extent that we (individually or collectively) make the moral commitment to submit our concepts about the social to the constraints imposed by our Constitution, we will our concept of (civil) marriage to be consistent with these. . . . We defer to the Court because, and to the extent that, we acknowledge the Court as the expert in the consistency of our legal concepts with the moral concepts embodied in constitutional provisions. (Mercier 2007)

This is the expertise of the Court: to establish whether a particular piece of legislation is consistent with other (more fundamental) legal concepts or principles. We can find the same reasoning at work in many landmark decisions made by constitutional courts all over the world: after a certain principle has been incorporated in the constitution (the right to live, die, give birth, speak, vote, run for office, marry, etc.) the citizens delegate to the courts all decisions concerning the consistency of particular bits of legislation with these higher principles. Such decisions are normative in the sense that they rely on the courts' expertise in the interpretation and application of fundamental political and moral principles, but also in the sense that they concern the regulation of people's behavior by means of formal norms incorporated in the legal code. What they do not do is resolve ontological issues concerning marriage, democracy, procreation, or life and death, qua social or natural kinds.

This is a very different view of normativity compared to the one endorsed by the ameliorative theorist. Normative factors do influence decisions concerning the identity of token institutions, of course, but indirectly and only insofar as the institutions are incorporated and regulated by the law. Normative considerations do not and should not influence ontological or scientific debates about institution-types, because legal, moral, and political experts have no epistemic authority in that domain. A judge who has been trained to interpret the law does not have the knowledge and expertise to determine whether a particular physiological state is a state of *biological* life or death, for example. Similarly, she is not able to adjudicate whether a certain contract is a specific instance of a more general type of institution like marriage (that's why judges seek the advice of expert witnesses like doctors or anthropologists). What a judge can do is to adjudicate whether the introduction of a certain concept in our legislation is compatible with the higher principles that are codified in the constitution.

Obviously it would be disingenuous to pretend that judicial decisions do not affect the way in which people represent the institution of marriage, or the terms they use when they talk about same-sex unions. Such decisions influence significantly both our social practices (how many gays and lesbians choose to marry) and our symbolic representations (whether we see same-sex unions as just another version of marriage or as a union of a completely different kind). If this was not the case, the heated controversies that surround such decisions would be incomprehensible. But legal decisions affect only the actual reference (what rules people follow here and now) and the stereotypes (the folk theories) associated with a social term. The *extension* of the term is not affected, in contrast, because it is determined by factors that no court has the power to change.

The effect of legal rulings interacts with current social practices and with people's normative convictions. After a legislative body has decided that a contractual relation between members of the same sex must be called "marriage," individual citizens are still entitled to believe and argue that heterosexual marriage is the only *legitimate* concept of marriage—that is, the only concept that ought to be recognized by the law. But if they do so, they must be ready to revise the constitutional principles that currently protect the citizens of their country against discrimination. Moreover, people are entitled to argue that heterosexual marriage is their preferred notion of marriage, or that it is the only notion of marriage that is consistent with their religious convictions. But neither the courts nor people's normative convictions determine what marriage is as a social institution. The claim that heterosexual marriage is the only kind of marriage *qua social kind* must be adjudicated by scientific experts who are able to assess the claim in light of the best theories and empirical data that are currently available.

REFERENCES AND FURTHER READINGS

The distinction between types and tokens of institutions can be found in an unpublished (and underappreciated) paper on money by Uskali Mäki (2004). The idea that biological species are individuals or particulars dates back to Ghiselin (1974) and Hull (1976). On marriage as a bundle of rights or rules that solve a cluster of related problems, see Leach's (1955) classic article. It is significant that this understanding of marriage emerged in anthropology when ethnographers discovered contractual arrangements that could not be easily assimilated to standard Western marriage, like polyandry and same-sex unions. On the latter, in particular, see Krige's (1974) influential discussion of "woman-marriage" in central Africa. The view that marriage predominantly regulates male-female relationships, incidentally, is a peculiar modern European idea. Since at least Lévi-Strauss, anthropologists have pointed out that in most cultures marriage institutions regulate relationships among males and among groups (families, clans). Mercier (2007) offers an excellent critique of semantic arguments about the nature of marriage, and a discussion of the scope of courts' decisions concerning institutional matters. Notice that the relation between legal rulings and folk conceptions is complex and multidirectional: sometimes the ruling changes the folk conception (as gay activists hope will happen in the case of marriage), and sometimes the folk conception or practice influences the law. In 1893 for example the US Supreme Court decided that tomatoes must be considered vegetables for all legal purposes concerning trading. With this ruling the folk concept of tomato was adopted as the legal concept, but, as Mercier points out, the Court did not and could not change the fact that, from a scientific point of view, a tomato is a fruit.

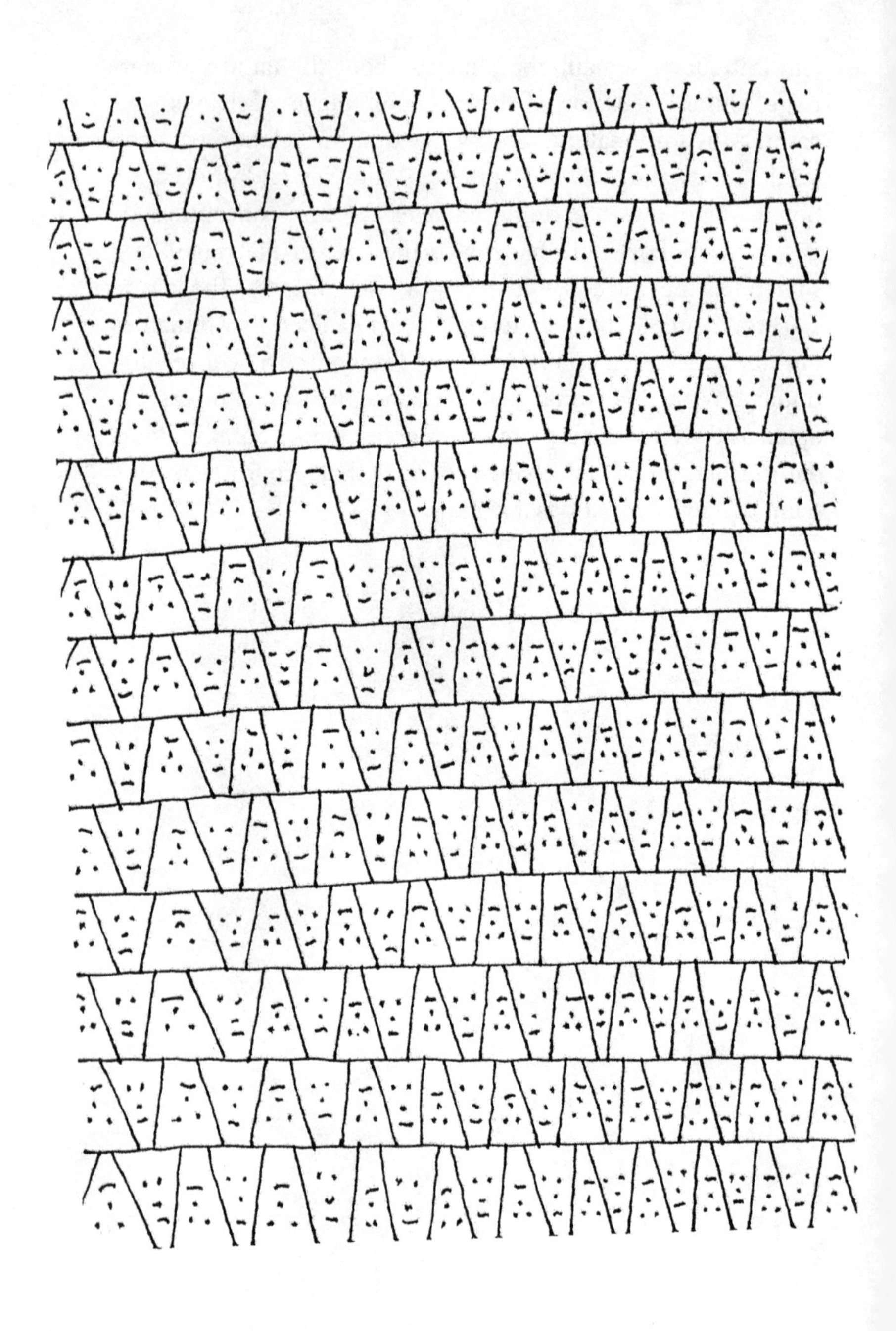

Bibliography

Aoki, M. (2001) *Toward a Comparative Institutional Analysis.* Cambridge, Mass.: MIT Press.

—— (2007) "Endogenizing Institutions and Institutional Change." *Journal of Institutional Economics* 3: 1–31.

—— (2011) "Institutions as Cognitive Media between Strategic Interactions and Individual Beliefs." *Journal of Economic Behavior and Organization* 79: 20–34.

—— (2015) "Why Is the Equilibrium Notion Essential for a Unified Institutional Theory? A Friendly Remark on the Article by Hindriks and Guala." *Journal of Institutional Economics* 11: 485–88.

Aumann, R. (1974) "Subjectivity and Correlation in Randomized Strategies." *Journal of Mathematical Economics* 1: 67–96.

—— (1976) "Agreeing to Disagree." *Annals of Statistics* 4: 1236–39.

—— (1987) "Correlated Equilibrium as an Expression of Bayesian Rationality." *Econometrica* 55: 1–18.

Austin, J. L. (1962) *Sense and Sensibilia.* Oxford: Oxford University Press.

Bacharach, M. (1999) "Interactive Team Reasoning: A Contribution to the Theory of Co-operation." *Research in Economics* 53: 117–47.

—— (2006) *Beyond Individual Choice.* Princeton: Princeton University Press.

Bardsley, N. (2007) "On Collective Intentions: Collective Action in Economics and Philosophy." *Synthese* 157: 141–59.

Barnes, S. B. (1983) "Social Life as Bootstrapped Induction." *Sociology* 17: 524–45.

—— (1989) "Ostensive Learning and Self-Referring Knowledge." In *Cognition and Social Worlds,* edited by D. Rogers and J. A. Sloboda. Oxford: Clarendon Press, pp. 190–204.

Bianchin, M. (2015) "Simulation and the We-Mode. A Cognitive Account of Plural First Persons." *Philosophy of the Social Sciences* 45: 442–61.

Bicchieri, C. (2001) *Rationality and Coordination.* Cambridge: Cambridge University Press.

—— (2006) *The Grammar of Society.* Cambridge: Cambridge University Press.

Binmore, K. (1994) *Game Theory and the Social Contract. Vol. 1: Playing Fair.* Cambridge, Mass.: MIT Press.

—— (1998) *Game Theory and the Social Contract. Vol. 2: Just Playing.* Cambridge, Mass.: MIT Press.

—— (2005) *Natural Justice*. Oxford: Oxford University Press.

—— (2008) "Do Conventions Need to Be Common Knowledge?" *Topoi* 27: 17–27.

—— (2010) "Game Theory and Institutions." *Journal of Comparative Economics* 38: 245–52.

—— (2015) "Institutions, Rules and Equilibria: A Commentary." *Journal of Institutional Economics* 11: 493–96.

Bloor, D. (1997) *Wittgenstein, Rules, and Institutions*. London: Routledge.

Bogen, J. (1988) "Comments on the Sociology of Science of Child Abuse." *Nous* 22: 65–66.

Boswell, J. (1994) *The Marriage of Likeness: Same-Sex Unions in Pre-modern Europe*. Oxford: Villard.

Boyd, R. (1983) "On the Current Status of the Issue of Scientific Realism." *Erkenntnis* 19: 45–90.

—— (1989) "What Realism Implies and What It Does Not." *Dialectica* 43: 5–29.

—— (1991) "Realism, Anti-foundationalism, and the Enthusiasm for Natural Kinds." *Philosophical Studies* 61: 127–48.

Boyer, P. and Petersen, M. B. (2012) "The Naturalness of (Many) Social Institutions: Evolved Cognition as Their Foundation." *Journal of Institutional Economics* 8: 1–25.

Brandom, R. (1994) *Making It Explicit*. Cambridge, Mass.: Harvard University Press.

Bratman, M. (1993) "Shared Intention." *Ethics* 104: 97–113.

—— (2014) *Shared Agency*. Oxford: Oxford University Press.

Broome, J. (2013) *Rationality Through Reasoning*. Oxford: Blackwell.

Buck, R. (1963) "Reflexive Predictions." *Philosophy of Science* 30: 359–69.

Calvert, R. L. (1998) "Rational Actors, Equilibrium, and Social Institutions." In *Explaining Social Institutions*, edited by J. Knight and I. Sened. Ann Arbor: University of Michigan Press, pp. 57–94.

Colingwood, R. G. (1946) *The Idea of History*. Oxford: Oxford University Press.

Conte, A. G. (1988) "An Essay on Constitutive Rules." *Poznan Studies in the Philosophy of the Sciences and the Humanities* 11: 251–57.

Cooper, R. (2004) "Why Hacking Is Wrong about Human Kinds." *British Journal for the Philosophy of Science* 55: 73–85.

Correia, F. (2008) "Ontological Dependence." *Philosophy Compass* 3: 1013–32.

Crawford, S. E. S. and Ostrom, E. (1995) "A Grammar of Institutions." *American Political Science Review* 89: 582–600.

Cubitt, R. and Sugden, R. (2003) "Common Knowledge, Salience and Convention: A Reconstruction of David Lewis' Game Theory." *Economics and Philosophy* 19: 175–210.

Daniels, N. (1996) *Justice and Justification: Reflective Equilibrium in Theory and Practice*. Cambridge: Cambridge University Press.

Devitt, M. and Sterelny, K. (1999) *Language and Reality*. Cambridge, Mass.: MIT University Press.

Dixit, A., Skeath, S., and Reiley, D. H. (2009) *Games of Strategy*. 3rd ed. New York: Norton.

Douglas, M. (1986) *How Institutions Think*. Syracuse: Syracuse University Press.

Dray, W. H. (1957) *Laws and Explanation in History*. Oxford: Oxford University Press.

Duncan, W. C. (2009) "Speaking Up for Marriage." *Harvard Journal of Law & Public Policy* 32: 915–30.

Dupré, J. (1993) *The Disorder of Things*. Cambridge, Mass.: Harvard University Press.

Elster, J. (1983) *Explaining Technical Change*. Cambridge: Cambridge University Press.

Epstein, B. (2015) *The Ant Trap: Rebuilding the Foundations of the Social Sciences*. Oxford: Oxford University Press.

Eskridge, W. N. (1993) "A History of Same-Sex Marriage." Faculty Scholarship Series, Paper 1504, Yale Law School.

Foucault, M. (1961/2006) *History of Madness*. London: Routledge.

—— (1979) *Discipline and Punish: The Birth of the Prison*. New York: Vintage.

Frank, R. (1987) *Passions within Reason*. New York: Norton.

Gadamer, H. G. (1960) *Truth and Method*. New York: Seabury.

Gallotti, M. and Frith, C. (2013) "Social Cognition in the We-mode." *Trends in Cognitive Sciences* 17: 160–65.

Gallotti, M. and Michael, J., eds. (2014) *Perspectives on Social Ontology and Social Cognition*. Dordrecht: Springer.

Gelman, S. A. (2003) *The Essential Child: Origins of Essentialism in Everyday Thought*. Oxford: Oxford University Press.

Ghiselin, M. T. (1974) "A Radical Solution to the Species Problem." *Systematic Zoology* 23: 536–44.

Gibbard, A. (1990) *Wise Choices, Apt Feelings*. Cambridge, Mass.: Harvard University Press.

Gilbert, M. (1989) *On Social Facts*. Princeton: Princeton University Press.

Gintis, H. (2007) "The Evolution of Private Property." *Journal of Economic Behavior and Organization* 64: 1–16.

—— (2009) *The Bounds of Reason*. Princeton: Princeton University Press.

Gold, N. (2012) "Team Reasoning, Framing and Cooperation." In *Evolution and Rationality*, edited by S. Okasha and K. Binmore. Cambridge: Cambridge University Press, pp. 185–212.

Gold, N. and Sugden, R. (2007a) "Collective Intentions and Team Agency." *Journal of Philosophy* 104: 109–37.

—— (2007b) "Theories of Team Agency." In *Rationality and Commitment*, edited by F. Peter and H. B. Schmidt. Oxford: Oxford University Press, pp. 280–312.

Goldman, A. I. (1989) "Interpretation Psychologized." *Mind and Language* 4: 161–85.

—— (2006) *Simulating Minds.* Oxford: Oxford University Press.
Goodhart, C. (1989) *Money, Information and Uncertainty.* Cambridge, Mass.: MIT Press.
Goodman, N. (1954) *Fact, Fiction, and Forecast.* Indianapolis: Bobbs, Merrill.
Gordon, R. (1986) "Folk Psychology as Simulation." *Mind and Language* 1: 158–71.
Greif, A. (2006) *Institutions and the Path to the Modern Economy.* Cambridge: Cambridge University Press.
Greif, A. and Kingston, C. (2011) "Institutions: Rules or Equilibria?" In *Political Economy of Institutions, Democracy and Voting*, edited by N. Schofield and G. Caballero. Berlin: Springer, pp. 13–43.
Griffiths, P. (2004) "Emotions as Natural and Normative Kinds." *Philosophy of Science* 71: 901–11.
Grünbaum, A. (1956) "Historical Determinism, Social Activism, and Predictions in the Social Sciences." *British Journal for the Philosophy of Science* 7: 236–40.
Grunberg, E. (1986) "Predictability and Reflexivity." *American Journal of Economics and Sociology* 45: 475–88.
Grunberg, E. and Modigliani, F. (1954) "The Predictability of Social Events." *Journal of Political Economy* 62: 465–78.
Guala, F. (2010) "Infallibilism and Human Kinds." *Philosophy of the Social Sciences* 40: 244–64.
—— (2013a) "The Normativity of Lewis Conventions." *Synthese* 190: 3107–22.
—— (2013b) "Reflexivity and Equilibria." *Journal of Economic Methodology* 20: 397–405.
—— (2014) "On the Nature of Social Kinds." In *Perspectives on Social Ontology and Social Cognition*, edited by M. Gallotti and J. Michael. Dordrecht: Springer, pp. 57–68.
Guala, F. and Hindriks, F. (2015) "A Unified Social Ontology." *Philosophical Quarterly* 65: 177–201.
Guala, F. and Mittone, L. (2010) "How History and Conventions Create Norms: An Experimental Study." *Journal of Economic Psychology* 31: 749–56.
Habermas, J. (1968/1971) *Knowledge and Human Interests.* Boston: Beacon.
Hacking, I. (1983) *Representing and Intervening.* Cambridge: Cambridge University Press.
—— (1986) "Making Up People." In *Reconstructing Individualism*, edited by P. Heller, M. Sosna, and D. Wellberry. Stanford: Stanford University Press, pp. 222–36; reprinted in Hacking (2002).
—— (1991) "A Tradition of Natural Kinds." *Philosophical Studies* 61: 109–26.
—— (1995) "The Looping Effect of Human Kinds." In *Causal Cognition: A Multidisciplinary Debate*, edited by A. Premack. Oxford: Clarendon Press, pp. 351–83.
—— (1999) *The Social Construction of What?* Cambridge, Mass.: Harvard University Press.

—— (2002) *Historical Ontology*. Cambridge, Mass.: Harvard University Press.

—— (2007a) "Kinds of People: Moving Targets." *Proceedings of the British Academy* 151: 285–318.

—— (2007b) "Natural Kinds: Rosy Dawn, Scholastic Twilight." *Royal Institute of Philosophy Supplement* 61: 203–39.

Hakli, R., Miller, K., and Tuomela, R. (2011) "Two Kinds of We-Reasoning." *Economics and Philosophy* 26: 291–320.

Haslanger, S. (1995) "Ontology and Social Construction." *Philosophical Topics* 23: 95–125; reprinted in Haslanger (2012).

—— (2012) *Resisting Reality: Social Construction and Social Critique*. Oxford: Oxford University Press.

Hayek, F. A. (1976) *Denationalization of Money*. London: Institute of Economic Affairs.

Hédoin, C. (2015) "Accounting for Constitutive Rules in Game Theory." *Journal of Economic Methodology* 22: 439–61.

Hindriks, F. (2005) "Rules and Institutions: Essays on Meaning, Speech Acts and Social Ontology." PhD dissertation, Erasmus University Rotterdam.

—— (2009) "Constitutive Rules, Language, and Ontology." *Erkenntnis* 71: 253–75.

Hindriks, F. and Guala, F. (2015a) "Institutions, Rules, and Equilibria: A Unified Theory." *Journal of Institutional Economics* 11: 459–80.

—— (2015b) "Understanding Institutions: Replies to Aoki, Binmore, Hodgson, Searle, Smith, and Sugden." *Journal of Institutional Economics* 11: 515–22.

Hodgson, G. M. (2006) "What Are Institutions?" *Journal of Economic Issues* 15: 1–23.

—— (2015) "On Defining Institutions: Rules versus Equilibria." *Journal of Institutional Economics* 11: 497–505.

Hofstadter, D. (1979) *Gödel, Escher, Bach: An Eternal Golden Braid*. New York: Basic Books.

Horwich, P. (1998) *Meaning*. Oxford: Oxford University Press.

Hull, D. L. (1976) "Are Species Really Individuals?" *Systematic Zoology* 25: 174–91.

Hume, D. (1748/2003) *A Treatise of Human Nature*. London: Penguin.

Hunter, N. D. (1991) "Marriage, Law, and Gender: A Feminist Inquiry." *Law & Sexuality* 1: 9–17.

Khalidi, M. A. (2010) "Interactive Kinds." *British Journal for the Philosophy of Science* 61: 335–60.

—— (2013) *Natural Categories and Human Kinds*. Cambridge: Cambridge University Press.

—— (2015) "Three Kinds of Social Kinds." *Philosophy and Phenomenological Research* 90: 96–112.

Kiyotaki, N. and Moore, J. (2001) "Evil Is the Root of All Money." The Clarendon Lectures (Lecture 1). http://www.princeton.edu/~kiyotaki/papers/Evilistherootofallmoney.pdf.

Kiyotaki, N. and Wright, R. (1989) "On Money as a Medium of Exchange." *Journal of Political Economy* 97: 927–54.

—— (1991) "A Contribution to the Pure Theory of Money." *Journal of Economic Theory* 53: 215–35.

Knapp, G. F. (1905/1973) *The State Theory of Money.* Clifton, N.J.: Augustus M. Kelley.

Knight, J. (1992) *Institutions and Social Conflict.* Cambridge: Cambridge University Press.

Knoblauch, H. (1996) "The Construction of Social Reality. By John R. Searle." *American Journal of Sociology* 101: 1459–61.

Koppelman, A. (1988) "The Miscegenation Analogy: Sodomy Law as Sex Discrimination." *Yale Law Journal* 98: 145–64.

Krige, E. J. (1974) "Woman-Marriage, with Special Reference to the Louedu—Its Significance for the Definition of Marriage." *Africa* 44: 11–37.

Kripke, S. (1972) *Naming and Necessity.* Cambridge, Mass.: Harvard University Press.

—— (1982) *Wittgenstein on Rules and Private Language.* Oxford: Blackwell.

Krishna, D. (1971) "The Self-Fulfilling Prophecy and the Nature of Society." *American Sociological Review* 36: 1104–7.

Kuhn, T. S. (1970) *The Structure of Scientific Revolutions.* Chicago: University of Chicago Press.

Kukla, A. (2000) *Social Constructivism and the Philosophy of Science.* London: Routledge.

Kuorikoski, J. and Pöyhönen, S. (2013) "Looping Kinds and Social Mechanisms." *Sociological Theory* 30: 187–205.

Kuran, T. (1995) *Private Truths, Public Lies.* Cambridge, Mass.: Harvard University Press.

Law, S. A. (1988) "Homosexuality and the Social Meaning of Gender." *Wisconsin Law Review* 187: 218–21, 230–33.

Leach, E. R. (1955) "Polyandry, Inheritance and the Definition of Marriage." *Man* 55: 182–86.

Lewis, D. K. (1969) *Convention: A Philosophical Study.* Cambridge, Mass.: Harvard University Press.

—— (1980) "Mad Pain and Martian Pain." In *Readings in the Philosophy of Psychology*, vol. 1, edited by N. Block. Cambridge, Mass.: Harvard University Press, pp. 216–22.

Luce, R. D. and Raiffa, H. (1957) *Games and Decisions.* New York: Wiley.

Machery, E. (2014) "Social Ontology and the Objection from Reification." In *Perspectives on Social Ontology and Social Cognition,* edited by M. Gallotti and J. Michael. Dordrecht: Springer, pp. 87–100.

Mäki, U. (2004) "Reflections on the Ontology of Money." Unpublished paper, Erasmus Institute for Philosophy of Economics.

—— (2012) "Realism and Antirealism about Economics." In *Philosophy of Eco-*

nomics. Handbook of the Philosophy of Science, vol. 13, edited by U. Mäki. Amsterdam: Elsevier, pp. 3–24.

Mallon, R. (2003) "Social Construction, Social Roles, and Stability." In *Socializing Metaphysics*, edited by F. F. Schmidt. Lanham, Md.: Rowman & Littlefield, pp. 327–53.

—— (2013) "Naturalistic Approaches to Social Construction." In *Stanford Encyclopedia of Philosophy*, edited by E. Zalta. http://plato.stanford.edu/entries/social-construction-naturalistic/.

Mantzavinos, C. (2001) *Individuals, Institutions, and Markets*. Cambridge: Cambridge University Press.

——, ed. (2009) *Philosophy of the Social Sciences*. Cambridge: Cambridge University Press.

Maynard Smith, J. (1982) *Evolution and the Theory of Games*. Cambridge: Cambridge University Press.

Menger, K. (1892) "On the Origin of Money." *Economic Journal* 2: 239–55.

Mercier, A. (2007) "Meaning and Necessity: Can Semantics Stop Same-Sex Marriage?" *Philosophical Essays* 8(1): article 14. http://commons.pacificu.edu/eip.

Merton, R. K. (1948) "The Self-Fulfilling Prophecy." *Antioch Review* 8: 193–210.

—— (1968) *Social Theory and Social Structure*. New Brunswick, N.J.: Transaction.

Michael, J. and Pacherie, E. (2015) "On Commitments and Other Uncertainty Reduction Tools in Joint Action." *Journal of Social Ontology* 1: 89–120.

Midgley, G. C. (1959) "Linguistic Rules." *Proceedings of the Aristotelian Society* 54: 271–90.

Mill, J. S. (1843) *A System of Logic*. London: Longmans, Green, Reader & Dyer.

Miller, S. (2010) *The Moral Foundations of Social Institutions*. Cambridge: Cambridge University Press.

Morgenstern, O. (1928) *Wirtschaftsprognose, eine Untersuchung ihrer Voraussetzungen und Möglichkeiten*. Vienna: Springer Verlag.

Morin, O. (2011) "Three Ways of Misunderstanding the Power of Rules." In *The Background of Social Reality*, edited by M. Schmitz, B. Kobow, and H.-B. Schmidt. Dordrecht: Springer, pp. 185–201.

Morton, A. (1994) "Game Theory and Knowledge by Simulation." *Ratio* 7: 14–25.

—— (2003) *The Importance of Being Understood*. London: Routledge.

Nagel, E. (1961) *The Structure of Science*. New York: Harcourt, Brace, and World.

Nichols, S. (2004) *Norms with Feelings*. Oxford: Oxford University Press.

North, D. (1990) *Institutions, Institutional Change and Economic Performance*. Cambridge: Cambridge University Press.

Osborne, M. J. (2004) *An Introduction to Game Theory*. Oxford: Oxford University Press.

Osborne, T. (1997) "The Limits of Ontology." *History of the Human Sciences* 10: 97–102.

Parsons, T. (1935) "The Place of Ultimate Values in Sociological Theory." *International Journal of Ethics* 45: 282–316.

Pettit, P. (1990) "Virtus Normativa: Rational Choice Perspectives." *Ethics* 100: 725–55.

—— (1996) "Functional Explanation and Virtual Selection." *British Journal for the Philosophy of Science* 47: 291–302.

Pickett, B. (2011) "Homosexuality." In *Stanford Encyclopedia of Philosophy*, edited by E. Zalta. http://plato.stanford.edu/entries/homosexuality/.

Polo, M. (1298/1974) *Il Milione.* Turin: Einaudi.

Pryor, F. L. (1977) "The Origins of Money." *Journal of Money, Credit and Banking* 9: 391–409.

Putnam, H. (1975) "The Meaning of 'Meaning.'" In *Mind, Language and Reality. Philosophical Papers*, vol. 2. Cambridge: Cambridge University Press, pp. 215–71.

Ransdell, J. (1971) "Constitutive Rules and Speech-Act Analysis." *Journal of Philosophy* 68: 385–99.

Rapp, C. (1997) "Ritual Brotherhood in Byzantium." *Traditio* 52: 285–326.

Rawls, J. (1955) "Two Concepts of Rules." *Philosophical Review* 64: 3–32.

—— (1971) *A Theory of Justice.* Oxford: Oxford University Press.

Raz, J. (1999) "Explaining Normativity: On Rationality and the Justification of Reason." *Ratio* 12: 354–79.

Ricoeur, P. (1965) *Interpretation Theory.* Fort Worth: Texas Christian University Press.

Rizzolatti, G. and Sinigaglia, C. (2008) *Mirrors in the Brain.* Oxford: Oxford University Press.

Romanos, G. D. (1973) "Reflexive Predictions." *Philosophy of Science* 40: 97–109.

Rosenthal, R. and Jacobson, L. (1968) *Pygmalion in the Classroom.* New York: Holt, Rinehart & Winston.

Rothbart, M. and M. Taylor (1992) "Category Labels and Social Reality: Do We View Social Categories as Natural Kinds?" In *Language, Interaction and Social Cognition*, edited by G. R. Semin and K. Fiedler. London: Sage, pp. 11–36.

Rousseau, J. J. (1755) *Discourse on the Origin of Inequality.* Dover Publications.

Roversi, C., Borghi, A. M., and L. Tummolini (2013) "A Marriage Is an Artefact and Not a Walk That We Take Together: An Experimental Study on the Categorization of Artefacts." *Review of Philosophy and Psychology* 4: 527–42.

Ruben, D.-H. (1989) "Realism in the Social Sciences." In *Dismantling Truth*, edited by H. Lawson and L. Appignanesi. London: Weidenfeld and Nicolson, pp. 58–75.

Schelling, T. (1960) *The Strategy of Conflict.* Cambridge, Mass.: Harvard University Press.

—— (1978) *Micromotives and Macrobehavior.* New York: Norton.
—— (2010) "Game Theory: A Practitioner's Approach." *Economics and Philosophy* 26: 27–46.
Schmitt, F. F., ed. (2003) *Socializing Metaphysics.* Lanham, Md.: Rowman & Littlefield.
Schotter, A. (1981) *The Economic Theory of Social Institutions.* Cambridge: Cambridge University Press.
Searle, J. R. (1969) *Speech Acts: An Essay in the Philosophy of Language.* Cambridge: Cambridge University Press.
—— (1990) "Collective Intentions and Actions." In *Intentions in Communication,* edited by P. Cohen, J. Morgan, and M. E. Pollack. Cambridge, Mass.: MIT Press, pp. 401–15.
—— (1995) *The Construction of Social Reality.* London: Penguin.
—— (2005) "What Is an Institution?" *Journal of Institutional Economics* 1: 1–22.
—— (2009) "Language and Social Ontology." In *Philosophy of the Social Sciences,* edited by C. Mantzavinos. Cambridge: Cambridge University Press, pp. 9–27.
—— (2010) *Making the Social World.* Oxford: Oxford University Press.
—— (2015) "On Status Functions and Institutional Facts: Reply to Hindriks and Guala." *Journal of Institutional Economics* 11: 507–14.
Sellars, W. (1963) "Imperatives, Intentions, and the Logic of 'Ought.'" In *Morality and the Language of Conduct,* edited by H. Castaneda and G. Nakhnikian. Detroit: Wayne State University Press, pp. 159–214.
Sillari, G. (2008) "Common Knowledge and Convention." *Topoi* 27: 29–39.
—— (2012) "Rule-Following and Coordination." *Synthese* 190: 871–90.
Simon, H. (1954) "Bandwagon and Underdog Effects and the Possibility of Election Predictions." *Public Opinion Quarterly* 18: 245–53; reprinted in *Models of Man* (New York: Wiley, 1957), pp. 455–557.
Skorupski, J. (2010) *The Domain of Reason.* Oxford: Oxford University Press.
Skyrms, B. (1996) *Evolution of the Social Contract.* Cambridge: Cambridge University Press.
—— (2004) *The Stag Hunt and the Evolution of Social Structure.* Cambridge: Cambridge University Press.
Smit, J. P., Buekens, F., and du Plessis, S. (2011) "What Is Money? An Alternative to Searle's Institutional Facts." *Economics and Philosophy* 27: 1–22.
—— (2014) "Developing the Incentivized Action View of Institutional Reality." *Synthese* 191: 1813–30.
Smith, V. L. (2015) "Conduct, Rules and the Origins of Institutions." *Journal of Institutional Economics* 11: 481–83.
Sober, E. (1983) "Equilibrium Explanation." *Philosophical Studies* 43: 201–10.
Sober, E. and Wilson, D. S. (1998) *Unto Others: The Evolution and Psychology of Unselfish* Behavior. Cambridge, Mass.: Harvard University Press.
Soros, G. (1987) *The Alchemy of Finance.* New York: Wiley.

—— (2013) "Fallibility, Reflexivity and the Human Uncertainty Principle." *Journal of Economic Methodology* 20: 309–29.

Sperber, D. (1996) *Explaining Culture.* Oxford: Blackwell.

Sugden, R. (1986/2004) *The Economics of Rights, Co-operation and Welfare.* 2nd ed. Oxford: Blackwell.

—— (1993) "Thinking as a Team: Toward an Explanation of Nonselfish Behavior." *Social Philosophy and Policy* 10: 69–89.

—— (1998a) "Normative Expectations: The Simultaneous Evolution of Institutions and Norms." In *Economics, Values, and Organization,* edited by A. Ben-Ner and L. Putterman. Cambridge: Cambridge University Press, pp. 73–100.

—— (1998b) "The Role of Inductive Reasoning in the Evolution of Conventions." *Law and Philosophy* 17: 377–410.

—— (2000a) "The Motivating Power of Expectations." In *Rationality, Rules, and Structure,* edited by J. Nida-Rümelin and W. Spohn. Dordrecht: Kluwer, pp. 103–29.

—— (2000b) "Team Preferences." *Economics and Philosophy* 16: 174–204.

—— (2003) "The Logic of Team Reasoning." *Philosophical Explorations* 6: 165–81.

—— (2015) "On 'Common Sense Ontology': A Comment on the Paper by Frank Hindriks and Francesco Guala." *Journal of Institutional Economics* 11: 489–92.

Tahko, T. E. and Lowe, E. J. (2015) "Ontological Dependence." In *The Stanford Encyclopedia of Philosophy,* edited by E. N. Zalta. http://plato.stanford.edu/archives/spr2015/entries/dependence-ontological/.

Tajfel, H. (1982) *Human Groups and Social Categories.* Cambridge: Cambridge University Press.

Tajfel, H. and Turner, J. C. (1986) "The Social Identity Theory of Intergroup Behaviour." In *Psychology of Intergroup Relations,* edited by S. Worchel and L. W. Austin. Chicago: Nelson-Hall, pp. 7–24.

Taylor, C. (1971) "Interpretation and the Sciences of Man." *Review of Metaphysics* 25: 3–51.

Thomas, W. I. (1931) "The Relation of Research to the Social Process." In *Essays on Research in the Social Sciences,* edited by L. S. Lyon, I. Lubin, L. Meriam, and P. G. Wright. Washington: Brookings Institution, pp. 175–94.

Thomasson, A. (1999) *Fiction and Metaphysics.* Cambridge: Cambridge University Press.

——. (2003) "Realism and Human Kinds." *Philosophy and Phenomenological Research* 68: 580–609.

Tieffenbach, E. (2010) "Searle and Menger on Money." *Philosophy of the Social Sciences* 40: 191–212.

Tummolini, L., Andrighetto, G., Castelfranchi, C., and Conte, R. (2013) "A Convention or (Tacit) Agreement betwixt Us: On Reliance and Its Normative Consequences." *Synthese* 190: 585–618.

Tuomela, R. (1995) *The Importance of Us.* Stanford: Stanford University Press.
—— (2002a) *The Philosophy of Social Practices.* Cambridge: Cambridge University Press.
—— (2002b) "Reply to Critics." In *Social Facts & Collective Intentionality,* edited by G. Meggle. Frankfurt: Hänsel-Hohenhausen AG, pp. 419–36.
—— (2007) *The Philosophy of Sociality.* Oxford: Oxford University Press.
Tuomela, R. and Miller, K. (1988) "We-intentions." *Philosophical Studies* 53: 367–89.
Turner, S. (1999) "Searle's Social Reality." *History and Theory* 38: 211–31.
—— (2010) *Explaining the Normative.* Cambridge: Polity Press.
Ullmann-Margalit, E. (1977) *The Emergence of Norms.* Oxford: Clarendon Press.
United States Conference of Catholic Bishops (2009) "Bishops Applaud Maine Vote Defending Marriage." News release, November 4.
Vanberg, V. J. (2008) "On the Economics of Moral Preferences." *American Journal of Economics and Sociology* 67: 605–28.
Vanderschraaf, P. (1995) "Convention as Correlated Equilibrium." *Erkenntnis* 42: 65–87.
—— (1998) "Knowledge, Equilibrium and Convention." *Erkenntnis* 49: 337–69.
—— (2001) *Learning and Coordination.* London: Routledge.
Vanderschraaf, P. and Sillari, G. (2013) "Common Knowledge." In *Stanford Encyclopedia of Philosophy,* edited by E. Zalta. http://stanford.library.usyd.edu.au/entries/common-knowledge/.
Venn, J. (1866) *The Logic of Chance.* London: Macmillan.
Vermaas, P. E. and Houkens, W. (2003) "Ascribing Functions to Technical Artefacts: A Challenge to Etiological Accounts of Functions." *British Journal for the Philosophy of Science* 54: 261–89.
Vromen, J. (2003) "Collective Intentionality, Evolutionary Biology and Social Reality." *Philosophical Explorations* 6: 251–65.
Weber, M. (1910) "Diskussionsrede zu dem Vortrag von A. Ploetz über Die Begriffe Rasse und Gesellschaft." In *Gesammelte Aufsätze zur Soziologie und Sozialpolitik.* Tübingen: Mohr, pp. 456–62.
—— (1922) *Economy and Society.* Berkeley: University of California Press.
Weisberg, M. (2006) "Water Is *Not* H_2O." *Boston Studies in the Philosophy of Science* 242: 337–45.
Whewell, W. (1847) *The Philosophy of Inductive Sciences, Founded upon Their History.* London: Parker.
Wilson, R. A. (1999) "Realism, Essence, and Kind: Resuscitating Species Essentialism." In *Species: New Interdisciplinary Essays,* edited by R. A. Wilson. Cambridge, Mass.: MIT Press, pp. 187–207.
Winch, P. (1958) *The Idea of a Social Science and Its Relation to Philosophy.* London: Routledge.
Wittgenstein, L. (1953) *Philosophical Investigations.* Oxford: Blackwell.
Wray, L. R. (1990) *Money and Credit in Capitalist Economies.* Aldershot: Elgar.

Index